SpringerBriefs in Space Life Sciences

Series Editors

Günter Ruyters
Markus Braun
Space Administration
German Aerospace Center (DLR)
Bonn, Germany

The extraordinary conditions of space, especially microgravity, are utilized for research in various disciplines of space life sciences. This research that should unravel – above all – the role of gravity for the origin, evolution, and future of life as well as for the development and orientation of organisms up to humans, has only become possible with the advent of (human) spaceflight some 50 years ago. Today, the focus in space life sciences is 1) on the acquisition of knowledge that leads to answers to fundamental scientific questions in gravitational and astrobiology, human physiology and operational medicine as well as 2) on generating applications based upon the results of space experiments and new developments e.g. in non-invasive medical diagnostics for the benefit of humans on Earth. The idea behind this series is to reach not only space experts, but also and above all scientists from various biological, biotechnological and medical fields, who can make use of the results found in space for their own research. SpringerBriefs in Space Life Sciences addresses professors, students and undergraduates in biology, biotechnology and human physiology, medical doctors, and laymen interested in space research. The Series is initiated and supervised by Dr. Günter Ruyters and Dr. Markus Braun from the German Aerospace Center (DLR). Since the German Space Life Sciences Program celebrated its 40th anniversary in 2012, it seemed an appropriate time to start summarizing – with the help of scientific experts from the various areas - the achievements of the program from the point of view of the German Aerospace Center (DLR) especially in its role as German Space Administration that defines and implements the space activities on behalf of the German government.

More information about this series at http://www.springer.com/series/11849

Jean-Pierre de Vera

Astrobiology on the
International Space Station

Jean-Pierre de Vera
Institute of Planetary Research
German Aerospace Center
Berlin, Berlin, Germany

ISSN 2196-5560 ISSN 2196-5579 (electronic)
SpringerBriefs in Space Life Sciences
ISBN 978-3-030-61690-8 ISBN 978-3-030-61691-5 (eBook)
https://doi.org/10.1007/978-3-030-61691-5

This Springer imprint is published by the registered company Springer Nature Switzerland AG.
The registered company address is: Gewerbestrasse 11, 6330 Cham, Switzerland

Foreword

"Astrobiology on the International Space Station" is the title of the present, the 12th, book of the series "Springer Briefs in Space Life Sciences." In order to deal with this fascinating topic of space research, the author Jean-Pierre de Vera has chosen a surprising, though interesting approach: instead of trying to provide an overview or a summary of astrobiological research and results of the last decades, he has taken the case study of the recently completed BIOMEX project to introduce the interested reader to major topics of astrobiology such as the search for life in the universe and the habitability of Mars and other heavenly bodies, thereby demonstrating the organizational hurdles, technical and scientific challenges, and accomplishments of astrobiological projects on the International Space Station. By this approach, the author is in the position to express his personal view on the various steps towards astrobiological experimentation in space, which may not be shared by all colleagues from the field, but will certainly lead to stimulating discussions. Together with three other international astrobiological projects, BIOMEX was realized in the timeframe 2014 to 2016 on the third ESA (European Space Agency) exposure facility EXPOSE-R2 located on the outside platform of the Russian Zvezda module of the International Space Station. Major BIOMEX objective was to investigate to what extent biomolecules such as biological pigments, cellular components, and biofilms are resistant to and able to maintain their stability under space and Mars-like conditions.

In the first of the two chapters of the book, the author provides detailed information on the ways to search for life on Mars by looking for biological molecules in space, the so-called biosignatures of life. Necessary pre-flight tests, ground reference experiments, and the spectroscopic analysis methods before and after long-term spaceflight are thoroughly described to set the scene for the space experiment as such. A description of significant results especially from Raman spectroscopy and confocal laser scanning microscopy follows. The chapter is concluded by a summary on Earth benefits of astrobiological research from the author's personal perspective—addressing in particular the stimulating effect on young scientists and the successful involvement of international students.

Chapter 2 focuses largely on the topic of habitability of Mars and the search of extant and/or extinct life on this planet. An important aspect here is research on the survivability of organisms of different domains of the tree of life. Again, the BIOMEX project on the EXPOSE-R2 facility outside ISS is taken as a case study for delivering important research results in astrobiology. In the end, a table on the habitability of Mars for various organisms is proving the impressive capability of survivability of different organisms. A very cautious statement on the lithopanspermia hypothesis—the idea that basic life forms can be distributed throughout the solar system via rock fragments cast forth by meteoroid impacts including the possibility of the origin of life on a distant planet with succeeding transfer of life to Earth—closes the book: "The likelihood of lithopanspermia in the solar system and specifically in the Earth-Mars trajectory is not zero."

Space Administration, German   Günter Ruyters
Aerospace Center (DLR)
Bonn, Germany
Space Administration, German   Markus Braun
Aerospace Center (DLR)
Bonn, Germany
August 2020

Contents

Chapter 1
Biomolecules in Space: The Way to Search for Life on Mars

Abstract This chapter will give a short overview on the first systematic analysis concept and first results on the resistance of biomolecules to space and Mars-like conditions realized on the exposure platform EXPOSE-R2 on the International Space Station (ISS). This platform in Low Earth Orbit (LEO) was chosen, because on ground even the best space and planetary simulation chambers are not able to cover all relevant environmental conditions, we might encounter directly in space or on Mars. A much better approach can only be realized directly in space and particularly on the long-term existing platform in Earth's orbit. It shows also the increased value the ISS has gained for astrobiological planetary research with its main topics of "habitability of planets," "limits of life," and "the search for life beyond Earth." The results presented here are mainly based on the successful space exposure experiment BIOMEX (*BIO*logy and *M*ars *EX*periment). The outcome of these and also previous and future space exposure experiments have direct implications for future space missions with the main goal to search for life in the solar system and beyond.

Keywords Biosignatures · Life detection · ISS · Mars · Low earth orbit · Raman spectroscopy

1.1 Introduction

The detection of life beyond Earth is significantly driven by the characterization of biomolecules, remnants of life such as fossils or traces obtained through bio mineralization as unambiguous biosignatures. These so-called biosignatures have even to be resistant to different extreme and sometimes diverging environmental conditions or they should be recognizable as such even after a certain degree of degradation. Otherwise we would not be able to detect those in an environment different to Earth. Because we only know one example for life in the universe, which means life on our home planet Earth, we have to be aware that all classifications and descriptions of life and its signatures are considered in an Earth-centric view. It means we are forced to use our planet Earth and its organisms as a reference system for the search for life in the universe. Besides getting a feeling on what kind of life forms we might expect to be present on or in some of the planetary bodies in our Solar System, also important tests on life detection instruments are needed. Those instrument tests will allow studying their suitability to operate and measure in extreme environments unambiguously biosignatures of extant or extinct life. It will also allow studying different signature patterns of characterized and selected biosignatures, which might occur, if those are measured under ambient terrestrial conditions or during and after exposure to extraterrestrial conditions. Protocols and collections on those data sets might help to analyze and interpret data correctly in future exploration missions with the main target to search for life beyond Earth (de Vera et al. 2019a).

The best platforms to get information on stability of biosignatures and the suitability of different life detection methods and instruments are planetary simulation labs and those much advanced facilities realized as exposure platforms in space, such as on satellites or the International Space Station (ISS, de Vera et al. 2012, 2019b; Rabbow et al. 2017; Rettberg et al. 2004). The advantage of space exposure platforms in Earth's orbit compared to the space/planetary simulation laboratories on ground are that all necessary space and planetary environmental conditions are simultaneously available whereas on ground just a few space and planetary environmental parameters could be reproduced in parallel (Rabbow et al. 2017; Rettberg et al. 2004). In particular the combination of short wavelength UV (ultraviolet), the complex cosmic ionizing radiation, temperature oscillations, micro-gravity, and low pressure is only available in space and provides a research environment for astrobiology that approaches very closely other solar system bodies and moons closer than laboratories on Earth can realize (de Vera et al. 2012; Rabbow et al. 2017).

1.2 Selection of Microorganisms and Biomolecules

The best approach to get a feeling about what kind of biosignatures we have to look for is to follow a systematic guideline (de Vera et al. 2019a). According to this guideline the first step is related to search for habitable niches in extreme terrestrial environments we might also expect on other planetary bodies in the solar system. This means, terrestrial environmental conditions, which were also approximately observed on, e.g. Mars (Hauber et al. 2019) or the icy ocean worlds of Jupiter and Saturn (Taubner et al. 2020; Dachwald et al. 2020) should be taken into consideration if we would like to look for extraterrestrial relevant samples of existing terrestrial life in such approaching extraterrestrial planetary environments on our own home planet. Thus, before starting experiments exposing any defined samples to conditions with planetary relevance in the lab and in space, we have to search for terrestrial planetary analogue field sites. Organisms, remnants of life or even fossils we could find in such kind of planetary analogue field sites will serve as excellent samples for further studies on their cell composition and biomolecules serving potentially as biosignatures, because they have adapted or were preserved in an environment very similar to what is observed on potentially life-relevant planets like

Mars or the icy moons of Jupiter and Saturn. But because the field sites are just approaching extraterrestrial planetary conditions, we have to be aware, that we do not know about the real capacity of the selected samples to also resist to very low atmospheric pressure or space vacuum and extreme irradiation fluxes, as it occurs on the planets and moons of interest.

1.2.1 Investigations in Planetary Analog Field Sites

In this chapter an example will be given, where and how we get biosignature and habitability test relevant samples with interest to be further investigated in space. A list of the finally selected biomolecules serving as potential biosignatures will be presented with reference to the scientific questions and goals (Sect. 1.2.2 and Table 1.1). For that, the ESA space exposure experiment BIOMEX (Biology and Mars Experiment) is an excellent example to demonstrate the entire logistic which is necessary to take into account before starting such kind of projects in space.

Table 1.1 List of selected biomarkers/biosignatures for further investigations in pre- and post-flight phase (adapted from de Vera 2020)

Biomolecule/complete organism		Description/role in nature	Biomarker Extant life	Biomarker Extinct life
(1) Complete organism	Methanogen archaeon	The organism is able to consume Mars atmospheric CO_2 and H_2 to produce methane	+	−
	Cyanobacterium	The organism is able to consume Mars atmospheric CO_2 and H_2O	+	−
(2) Key-molecules for cell structure, physiology, protective role	Lipids	Main components of cell membranes and vesicles, limit to surrounding environment, frontier necessary for redox potentials	+	+
	Bacterial cellulose	Ingredients of cyanobacteria, alga, and plant cell wall	+	±
	Melanin	An important photo-protective role, possible energy harvesting pigments	+	−
	Carotene	excellent antioxidant properties/ resistance to UV irradiation, membrane stabilization, accessory pigments for the photosynthesis apparatus	+	±
(3) By-products, biominerals	Calcium carbonate	By-product, sometimes shielding against high irradiation	±	±
	Whelwellite	By-product, with distinct function to balance the water uptake and gas exchange	+	+

According to the previous descriptions on which samples have to be collected, it is clear, that those samples could only be found in the most extreme terrestrial environments, as there are deserts, alpine areas or Polar Regions. Therefore it is necessary to find the right logistic partners which are usually operating in such kind of very remote environments. In Germany those partners for terrestrial polar and ocean environments are mainly the Alfred Wegner Institute (AWI), the Federal Institute for Geosciences and Natural Resources (BGR), the Center for Marine Environmental Science (MARUM), and the Helmholtz Center for Ocean Research GEOMAR. In what it concerns the BIOMEX project, AWI and mainly the BGR supported the BIOMEX-team particularly by geologic and logistic operations in Antarctica but also by providing samples from the Arctic. Operations done in those planetary analogue environments are realized by ships, boats, all-terrain vehicles, snowmobiles, planes, and helicopters (Fig. 1.1).

Significant efforts of persuasion are needed to get the necessary support and prepare the right needed logistics. This could be achieved, if a more remote overview on the terrestrial areas of interest is deeply investigated largely in advance and if this is shown to the logistic partners for evaluation of the feasibility of the planned enterprises. A number of available maps at BGR but also from Earth observation satellites of DLR (TanDEM-X DEM), Digital Globe Foundation, USGS EROS Data Center, data available from the US Geological Survey were studied before proposing the project to the potential logistic partners and starting the necessary expeditions. The photographs and cartography maps of mainly permafrost areas with relevance to Mars were checked on Mars-analog geological/geomorphological similarities which are indicating the activity of water even under subzero temperatures, what is most important for the search for present life living in these extreme harsh environments because of the available liquid water amount. Those geo-bio-relevant features could be easily be seen also from orbit by analyzing the obtained images on the presence of Mars-analog gullies, alluvial fans, debris flows, shallow subsurface pathways or the appearance of permafrost structures such as polygons (Fig. 1.2).

As soon as it becomes clear what kind of logistic is needed, further reconnaissance flights with planes and helicopters could be planned, for further selection of the right Mars-analog field sites and finally landing in these areas of interest for deep in situ investigations on past and present water activity, on measuring micro-climate data such as temperature, humidity, and irradiation for the classification of the habitability of the selected field site and, finally, the search for the presence or absence of life. Organisms which were finally found in these planetary analog environments are potential candidates to be used for further analysis to planetary/Mars-simulations in the lab and finally in space. A first check on the physiological activity of organisms in their own habitat in such terrestrial planetary analog environments even could show, if the finally selected and collected organisms have originally also colonized the investigated and classified planetary analog field sites and are living there really in their own habitat and not blown by wind occasionally in this areas of interest (see also de Vera et al. 2014; Schulze-Makuch et al. 2018).

Fig. 1.1 Logistic needed in the pre-flight phase for planetary analog field site investigations in North Victoria Land/Antarctica needed for the pre-selection of samples to be used in space for the BIOMEX space experiment: helicopter, snowmobiles, planes, and ships (photos taken during GANOVEX X, XI, and XIII)

1.2.2   *Finally Selected Biomolecules Serving as Potential Biosignatures*

An overview on the selected organisms on the planetary analog field sites will be shown later in Sect. 2.2. This paragraph is focusing on the selection of biomolecules as potential biosignatures to be used as reference for future exploration missions with the main goal to search for life on Mars. The selection of these molecules was driven

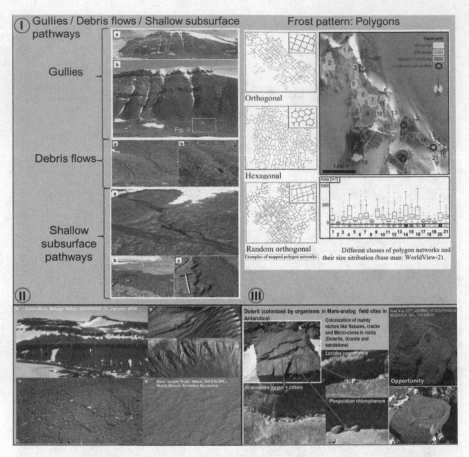

Fig. 1.2 Mars-analog field sites in North Victoria Land (Antarctica): (**a**) Mars-analog water erosion features; (**b**) Mars-analog gullies on slopes and polygons; (**c**) Colonized Mars-analog rocks: sandstone and dolerites

by its abundant presence in a diversity of Early Earth-, Mars-, and planetary-relevant organisms (mainly also present in the collected organisms) as well as its specific spectroscopically detectable characteristics. It has to be emphasized, that the final decision, if those pre-selected biomolecules are also really suitable to be used as biosignatures could be done after running the tests performed in the laboratories and in space (see Sects. 1.5, 1.6, and 1.7).

More or less complex organic molecules which are the ingredients of cells or which are produced by any kind of life forms are defined from an astrobiological point of view as biomarkers or biosignatures (de Vera 2020). But also the interaction with the environment has to be taken into account. Fossilization (Cady et al. 2003), biogenic mineral transformation or bleaching (Gadd 2007; Weber et al. 2012) as well as the consumption or releases of gases are considered as biosignatures (de Vera 2020).

In general a biosignature can be considered as a fingerprint of present or past life (de Vera 2020). Microorganisms often leave various traces as detectable evidence of their presence. These bio-traces can be detectable even long after been dead or extinct (Boston et al. 2004). In addition to microfossils, microbial mats and biofilms are capable of producing biofabrics in rocks and large-scale structures such as microbialites that are biological in origin (Westall 1999; Allen et al. 2000). We can conclude that different kind of biosignatures could be possible. In general we can distinguish the following different biosignature classes (de Vera 2020):

1. a group of molecules which are forming the tiniest entity of a living or dead cell,
2. biominerals as mineral deposition produced through life forms (Boston et al. 2004; Banfield et al. 2004),
3. bio-leaching (Navarrete et al. 2013),
4. gaseous byproducts of metabolism what is detectable in atmospheres or dissolved in water (de Vera 2020),
5. fossils (Westall 1999).

Here are some examples for biomolecules, group of biomolecules, and even complete organisms that were chosen for the first space experiment with a focus on this specific topic of "life detection" and the reason why this detectable bio-material is interesting to be further analyzed and used as reference biosignature for the search for life on, e.g. Mars (see also Table 1.1):

1. Complete organism: e.g. Mars-relevant methanogen archaea or cyanobacteria,
2. key-molecules with stabilizing quality of the cell structure or with a physiological or protective role in microorganisms, such as lipids, carotene, bacterial cellulose, chlorophyll, melanin,
3. molecules as by-products of physiological activity such as $CaCO_3$ and whewellite $(Ca(C_2O_4) \cdot H_2O)$.

To (1): a complete organism is a direct proof of the presence of extant life and could be unambiguously identified as a direct biosignature. Besides the bacteria, methanogen archaea are predestinated to be organisms which originated and evolved very early in Earth's history and are also today mainly part of the extremophiles dominating habitats with extreme environmental conditions (Serrano et al. 2015; de Vera 2020). If life arose once on the planet Mars or even on the icy moons of Jupiter and Saturn, those organisms could be model organisms to study and particularly to investigate their cell components as potential biosignatures, if they are detectable by the current detection technology. Even the production of methane by those organisms has a direct link to the controversially and highly debated observations made on Mars, which seem to indicate the appearance and disappearance of methane in the Martian atmosphere (Formisano et al. 2004; Mumma et al. 2009; Webster et al. 2015; Lefèvre and Forget 2009; Zahnle et al. 2011).

Another group of organisms which could be either detected as a complete organism by Raman spectroscopy or fluorescence spectroscopy are cyanobacteria (Baqué et al. 2015, 2018, 2020; Billi et al. 2019). Cyanobacteria were chosen as candidates for potential extinct or even extant life forms on Mars (Böttger et al.

2012), because like other prokaryotes (archaea and bacteria) they appeared on early Earth at least 3.8–3.5 billion years ago (Gya) (Stackebrandt 2004). The earliest microfossils found on Earth (Marshall et al. 2011) was discovered in the Apex chert of the Warrawoona Group, Australia, dating from about 3.5 billion years ago and resemble coccoid (spherical) and filamentous cyanobacteria (Schopf 1993). Although there is still a debate on the interpretation of these findings, results from Brasier et al. (2002) were later confirmed with the conclusion that microfossils of prokaryotes were present in the same area of Australia as Schopf and his group had previously analyzed (Wacey et al. 2011).

To (2): complex macro-molecules which are forming and stabilizing cell walls or the cell membranes and are in direct contact to the surrounding solid, gaseous or liquid medium are also the first molecules which would be in contact to the detecting technologies such as laser scans and following spectroscopic investigations. Therefore, cell wall components such as cellulose or cell membrane components such as lipids are of high importance to be studied on their stability and the reconnaissance ability by the used detection technologies after exposure to simulated planetary conditions or direct space conditions. In respect of the membrane components also photo-protective pigments or physiologically interesting pigments should be taken into account.

Referring to the class of photo-protective molecules melanin is a good candidate (Pacelli et al. 2019). Melanins are promising biosignature candidates; they are distributed among all the kingdoms of life such as archaea, bacteria, and eukaryotes. And even if synthesized through different pathways it leads to suggest an early emergence in the evolution of life. Melanins confer on organisms a number of useful characteristics that allow them to adapt to extreme environments (Pacelli et al. 2019). Melanins also have unique physical, chemical, paramagnetic, and semiconductive properties (Meredith and Sarna 2006) that act as possible energy harvesting pigments (Dadachova et al. 2007; Gessler et al. 2014). This double role of protection and energy transduction may have been crucial for microbes in the early history of life on Earth when radiation was significantly higher than today (Pacelli et al. 2019).

Another pigment molecule which has a multitude of functions and is relevant as well to photo-protection as to physiologic reactions allowing the important energy gaining redox potential at membranes is carotene and its carotene derivates as part of the family of carotenoids. Their early role in membrane stabilization, prior to fatty acids, has been proposed (Ourisson and Nakatani 1994). Carotenes have also excellent antioxidant properties (Stahl and Sies 2003), what is essential for the organisms to resist, e.g. the intensely UV irradiated early Earth (Cockell 2002; Rothschild 1990). This molecule is widely distributed in the tree of life and is present in its three domains of life such as bacteria, archaea, and eukaryotes. Here the results on the carotene derivate Deinoxanthin embedded in the cell of *Deinococcus radiodurance* will be presented (Leuko et al. 2017).

A typical cell wall forming molecule is cellulose. The bacterial cellulose is a widespread biopolymer-molecule which is able to form three-dimensional matrices with an important role for the formation of biofilms (Zaets et al. 2014; Ross et al. 1991). The cellulose biosynthesis is very wide spread. It is known from plants, algae, metazoans, fungi, cellular slime molds, and bacteria (Ross et al. 1991; Okuda et al. 2004; Grenville-Briggs et al. 2008; Kawano et al. 2011). Cellulose could have played also an important role in the formation of microbial mats in pristine ecosystems on Earth even 3.5 billion years ago (Nobles et al. 2001; Nobles and Brown 2004). As an old microbial polymeric substance and matrix of some biofilms, it could also be better preserved as other molecules (de Vera 2020). According to Westall et al. 2000, fossilized microbial polymeric substances and biofilms are better preserved than fossil bacteria themselves. Such an old molecule could therefore also have relevance for Mars and analysis on the detectability of cellulose is necessary (de Vera 2020).

To (3) molecules as resulting by-products of physiological reactions forming excreted bio-mineralized structures could serve if remained over longer time scales as typical (micro-) fossils. Because Calcium Carbonate ($CaCO_3$) is produced by methanogen archaea species (Serrano et al. 2014) but is not an unambiguous biosignature and could also be produced abiotically, the bio-mineral whelwellite ($Ca(C_2O_4) \cdot H_2O$), could be a promising molecule to be further studied after exposure to space conditions (Böttger et al. 2014; de Vera 2020). This bio-mineral is assumed to balance the water uptake and gas exchange during the rare, physiologically favorable moist to wet environmental periods (Böttger et al. 2014). Calcium oxalates are rarely formed geologically as, e.g. reported from analysis of hydrothermal deposits but their biogenic formation mainly in coal and sedimentary nodules is much more frequent (Hoefs 1969; Zák and Skála 1993; Vassilev and Vassilev 1996; Ward 2002) and are therefore much more an indicator for the presence of extant life or the presence of former existing life (extinct life) in an investigated environment. Several extremotolerant microorganisms were identified to excrete oxalic acid into their surrounding where it disintegrates the rock and binds leaching calcium ions to form finally calcium oxalate deposits (Sohrabi 2012). But the extraction and isolation of this molecule for testing it directly to the space conditions was not trivial and therefore this molecule was finally exposed attached to its originating lichen *Circinaria gyrosa*. Results on this molecule after space exposure are still pending.

All detection tests of all mentioned selected biosignatures above were performed during the pre-flight and post-flight phase with Raman spectroscopy to check on one hand the suitability of this method for life detection operations and on the other hand to get information about spectral signatures of these marker-molecules. The Raman spectroscopy was chosen, because this method is foreseen to be used in the future ESA space experiment ExoMars (Edwards et al. 2012, 2013), and in further missions to Mars such as the Mars 2020 rover (Beegle et al. 2015) as well as potentially for the icy moons in the solar system (Dachwald et al. 2020).

1.3 Sample Preparation

Blank RLS spectra of pure biogenic compounds were used for preliminary characterization and as references for the subsequent studies. Therefore, all compounds were suspended in methanol (p. a.), dispersed on quartz glass slides, and the methanol was allowed to evaporate before Raman analysis. For the pre-flight sample verification test (SVT) at MUSC at DLR Cologne (see paragraphs below) and the BIOMEX mission itself, a set of samples was dispersed on round ⌀ 6 mm quartz discs (provided by RUAG, Spain).

These samples allow insight into the degradation of pure biogenic substances during simulation and actual space exposure. Moreover, the selected biogenic compounds were mixed with Sulfatic and Phyllosilicatic Mars Regolith Simulant powders of 25–1000 μm grain size (S-MRS and P-MRS, compositions in Table 1.2) at 5 % (w/w) to enable complete interaction between the compounds, potentially degrading Martian substrates, and—during exposure or simulation—the applied environmental conditions. In order to facilitate safe handling, to grant sufficient amounts during SVT simulation or LEO-exposure of the BIOMEX samples, to provide a smooth surface for RLS analysis, and, thus, to reduce multiple scattering, the mixture was pressed into ⌀ 6 mm pellets of 0.4 g each at 6 t for 15 min in a PP-10 pellet press (Retsch, Germany). The procedure was used for all samples analyzed before and after the implementation of the pre-flight SVT but also for preparation of the BIOMEX space exposure samples itself. All samples were kept dark, dry, and cold (+4 °C) between production, exposure, and analysis. RLS spectra of the MRS mineral compounds were published in Böttger et al. (2012) (Fig. 1.3).

Table 1.2 Composition of Martian Regolith simulants (MRS)

Sulfatic MRS		Phyllosilicatic MRS	
Mineral	Weight (%)	Mineral	Weight (%)
Gabbro	32	Montmorillonite	45
Gypsum	30	Chamosite	20
Dunite	15	Quartz	10
Hematite	13	Iron(III)-oxide	5
Goethite	7	Kaolinite	5
Quartz	3	Siderite	5
		Hydromagnesite	5
		Gabbro	3
		Dunite	2

Composition of sulfatic and phyllosilicatic Mars Regolith Simulant (S-MRS and P-MRS) as percentage of mineral net weight

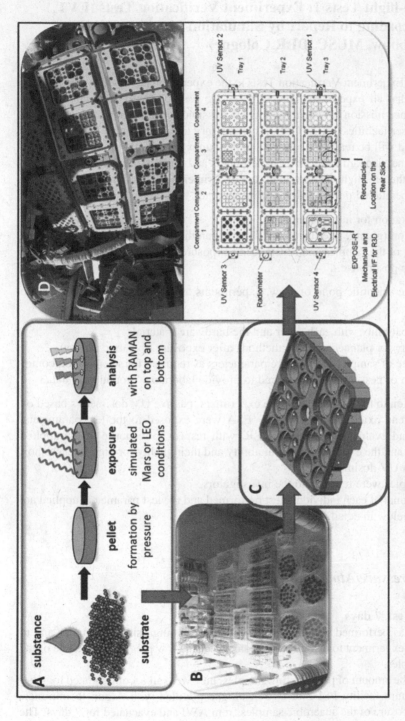

Fig. 1.3 Mars-analog mineral mixtures pressed to pellets after adding biomolecule substances. (**a**) Sample preparation: biomolecule droplets were added to Mars-analog mineral mixture powder and pressed to pellets. Those pellets were ready for exposure experiments in the Lab and in space. (**b**) Overview on samples prepared for pre-flight and flight experiments. (**c**) Sample holder compartment hardware. (**d**) EXPOSE-R2 platform: top on ISS, bottom sketch (adapted figure; Cottin et al. 2017)

1.4 Pre-flight Tests 1: Experiment Verification Tests (EVT, According to Report by Simulation Manager Dr. Elke Rabbow, MUSC, DLR Cologne)

The aim of Experiment Verification Tests in the experiment pre-flight test program is to ensure that all experiments and samples are well prepared and suited for the planned space mission to obtain the highest possible scientific output. In particular for multi-user facilities as EXPOSE, it is necessary to ensure that all experiments fit together and will be ready for integration. Any unexpected failure of an experiment or sample may jeopardize all other experiments and the mission success.

During the pre-flight test program, the experiments and samples were analyzed from the operational point of view with respect to

- transportation for integration,
- integration (gluing, size (hardware fitting), handling),
- integrity of the samples and resistance to exposure parameters (integrity of form, size, composure, etc.).

From the scientific point of view, experiments and samples are investigated to ensure

- biocompatibility with each other and the hardware material,
- suitability for planned analysis methods after exposure (dynamic ranges),
- resistance of samples to exposure parameters as required for scientific outcome,
- selection of best fitting samples and to provide laboratory ground base data.

In addition to the samples from the experiment, passive UV dosimeters based on polyphenylene oxide and provided by ESA were exposed to the test parameters, analyzed and evaluated at MUSC, DLR, with respect to response to the applied parameters and the expected space suitability and their quality for high fluence short wavelength UV dosimetry.

All samples were returned to the investigators.

The rational of each individual test performed and the test parameters applied are described below in detail.

1.4.1 Pressure/Atmosphere

Vacuum Test, 7 days

This test was performed to detect any problems with outgassing of samples and to provide an experiment to examine the desiccation and low pressure resistance of the tested samples.

Due to the amount of participating samples, they were all accommodated together in the vacuum facility PSI 2 at DLR Cologne prefilled with Argon to minimize oxygen exposure of the anaerobic samples from AWI and evacuated for 7 days. The final pressure reached was 3.5×10^{-2} Pa, indicating some outgassing problems. After evacuation, the facility was flooded with Argon inert gas, opened and the samples taken out immediately.

The empty facility tested directly after the run and reached a final pressure of 4×10^{-4} Pa in 24 h.

1.4.2 Mars atmosphere (CO_2 gas composition) 10^3 Pa pressure, 7 days

This test was performed to provide a parallel comparison experiment to the 1 week vacuum test to examine the resistance of the tested samples to the simulated Martian pressure of 650 Pa of a simulated Mars atmosphere composed of 95.55% CO_2, 2.70% N_2, 1.60% Ar, 0.15 O_2, ~370 ppm H_2O, provided pre-mixed by Praxair Deutschland GmbH.

All participating samples were accommodated together in the vacuum facility PSI 2. As for the vacuum test, the facility was flooded with Argon. After sample accommodation, the inner atmosphere was exchanged by simultaneous evacuation and flooding of the facility with the mixed gas at approximately ambient pressure. Pressure was reduced by stopping the gas flooding and evacuating the facility until a final pressure of 650 Pa was reached. The vacuum pump was disconnected from the facility and pressure was monitored. The final pressure did not change during the exposure. The facility was flooded after 1 week with Argon, opened and the samples taken out immediately.

1.4.3 Temperature

Temperature Cycles −10 °C to +45 °C, 48 Cycles
This test was performed to expose the samples to temperature cycles through the freezing point of 0 °C several times and to upper and lower temperatures frequently experienced during the previous two EXPOSE missions and expected during the EXPOSE-R2 mission.

All participating samples were accommodated together in the temperature controlled vacuum facility PSI 1 flooded with Argon at ambient pressure by displacement of air to avoid water precipitation on the samples during the cold phases. Temperature cycles were programmed with temperature monitoring by sensors attached to the sample carriers and feedback mechanism. Temperature was programmed to perform a complete temperature cycle in 8 h and though a total of 40 cycles were anticipated (because of margins for all sample integration and de-integration) it was possible to perform 48 cycles in the time period available.

Temperature was monitored and the samples regularly inspected through the top quartz window of PSI 1. As expected, the Argon prevented any water precipitation.

The program was stopped after the 49th cooling, and the facility given time to equilibrate with RT before it was opened and samples were taken out immediately.

Temperature Minimum −25 °C, 1 h
This test was performed to examine the resistance of the samples to the temperature extreme of −25 °C, as it was nearly experienced during the recent EXPOSE missions and may occur for EXPOSE-R2.

All participating samples were accommodated together in the temperature controlled vacuum facility MaSimKa. Before closure of the facility, it was flooded with Argon at ambient pressure by displacement of air to avoid water precipitation on the samples during the cold phase. Temperature was monitored by sensors attached to the sample carriers. After 50 min of cooling, samples were exposed to −25 °C ± 0.5 °C for 1 h before the facility was allowed to equilibrate to RT, opened and the samples were taken out immediately.

Temperature Maximum +60 °C, 1 h
This test was performed to examine the resistance of the samples to the temperature extreme of + 60°C, as it was experienced during the recent EXPOSE missions for less than 1 h and may occur for EXPOSE-R2.

All participating samples were accommodated together in a temperature controlled incubator. Temperature was monitored by sensors attached to the sample carriers. Samples were exposed to +60 °C ± 0.5 °C for 1 h before the facility was allowed to equilibrate to RT, opened and the samples were taken out immediately.

1.4.4   UV Radiation

Monochromatic Irradiation with UV-C
This test was performed to provide standardized UVC irradiation fluences by a mercury low pressure lamp in an UV-laboratory to obtain fluence-effect curves for the participating samples. Obtained data support a more precise estimation of UV resistance of the samples and for the selection of appropriate neutral density filters for the flight experiment.

The spectrum of the Hg low pressure lamp was determined with a calibrated Bentham150 spectroradiometer.

All irradiation samples of each individual experiment were accommodated together in the homogeneously irradiated field under the mercury low pressure lamp in closed sample carriers to minimize contamination risk and only opened for the irradiation (photos were also mainly taken with sample carriers like multiwell plates still closed) (Table 1.3).

Irradiance was measured with a calibrated UVX Meter, UVP, resulting in 56 µW cm^{-2} at the sample site. Stability of irradiance was confirmed before and

Table 1.3 Experiment verification tests (EVT)

EXPOSE-R2 EVT part1	BIOMEX experiment
Test parameter	Performed
Vacuum 10^{-5} Pa	7 day, pressure: $3.5 \times 10^{-2} \pm 0.12$ Pa
Mars atmosphere (CO_2 gas composition) 103 Pa	7 day, pressure: $6.5 \times 10^2 \pm 0.12$ Pa
Temperature $-10\,°C$ to $+45\,°C$	48 cycles 8 h each
Temperature max and min $-25\,°C$ and $+60\,°C$	$-25\,°C \pm 0.5\,°C$, 1 h $+60\,°C \pm 0.5\,°C$, 1 h
Irradiation 254 nm Hg low pressure lamp @ 56 µW/cm^2	$0\ s \rightarrow 0$ J/m^2 $18\ s \rightarrow 10.1$ J/m^2 $2\ min\ 59\ s \rightarrow 100.2$ J/m^2 $29\ min\ 46\ s \rightarrow 1000.2$ J/m^2 $4\ h\ 57\ min\ 37\ s \rightarrow 9999.9$ J/m^2
EXPOSE-R2 EVT part 2 (run 1 + 2)	BIOMEX experiment
Run 1 Irradiation 200–400 nm SOL2000 @1271.2 W/m^2 $_{200-400nm}$	$0\ s \rightarrow$ dark $18\ min \rightarrow 1.4 \times 10^3$ kJ/m^2 $3\ h \rightarrow 1.4 \times 10^4$ kJ/m^2 $30\ h \rightarrow 1.4 \times 10^5$ kJ/m^2 $99\ h \rightarrow 4.5 \times 10^5$ kJ/m^2 $148\ h \rightarrow 6.8 \times 10^5$ kJ/m^2
Run 2 Irradiation 200–400 nm SOL2000 @1271.2 W/m^2 $_{200-400nm}$ (as for a 12 month mission duration)	$0\ s \rightarrow$ dark $432\ s \rightarrow 5.5 \times 10^2$ kJ/m^2 (0.1 % ND filter) $1\ h\ 12\ min \rightarrow 5.5 \times 10^3$ kJ/m^2 (1.0 % ND filter) $30\ h \rightarrow 1.4 \times 10^5$ kJ/m^2 $60\ h \rightarrow 2.7 \times 10^5$ kJ/m^2 $120\ h \rightarrow 5.5 \times 10^5$ kJ/m^2
Gluing test	>24 h vulcanization, glue: Wacker-silicone

after each irradiation. For irradiation with the required fluences, samples were covered light tight with aluminum plates after the appropriate irradiation times. For practical reasons, irradiation times were in full seconds. Final fluences were 0, 10.1, 100.2, 1000.2, and 9999.9 J/m^2.

1.4.5 Gluing Test

This test was performed to investigate if the provided samples can be glued with the space approved non-outgassing glue Wacker-silicone RTV-S 691 A + B to an aluminum carrier, if they stick and if they can be removed again without destroying the samples and to investigate the biocompatibility of the glue with the EXPOSE-R2 biological samples. For each sample, a drop of the ready mixed glue was deposited onto a horizontal metallic surface and the sample attached. After vulcanization of the glue for 24 h, the metallic plate with the samples attached was brought into a vertical position, then turned head down and finally smashed forcefully in the vertical position with one edge onto a surface to test if the samples all stick.

Afterwards, all samples were removed by carefully applying sheering forces as it would be for sample de-integration after flight.

1.4.6   Control

For the irradiation experiment, additional 0 J m^{-2} "dark" samples were requested. These samples were transported and stored as all others, but not exposed. They represent the control for the irradiation test, but also for all other test of the EXPOSER2 EVT part 1 run 1 and run 2.

1.5   Pre-flight Tests 2: Scientific Verification Tests (SVT)

The SVT is designed to assure suitable sample preparation, and to test pre-flight integration as well as post-flight de-integration procedures at the EXPOSE-R2 hardware. Therefore, the samples were transferred to MUSC and integrated into the almost identical EXPOSE-R hardware (Fig. 1.4). Moreover, the SVT simulated mission equivalent space conditions to test the samples' resilience to the extreme non-terrestrial conditions during the BIOMEX mission. In analogy to BIOMEX itself, the SVT samples were integrated into tray 2, compartment 2 of the hardware and a broad range of Martian parameters was simulated for a period of 38 days: subzero temperature of -23 °C, Martian atmosphere composed of 95.55% CO_2, 2.70% N_2, 1.60% Ar, 0.15% O_2, and ~370 ppm H_2O (Praxair Deutschland GmbH) as well as a Mars-analogue pressure of 780–930 Pa. The upper layer of the three stacked sample sets (refer to Fig. 1.4b) was irradiated with UVR simulating a mission period of 12 months. The solar simulator SOL2000 irradiated the samples with an irradiance of 1271 W m^{-2} for an accumulated period of 5924 min leading to a total fluence of $5.7 \times 10^5 \text{ kJ m}^{-2}$ at $\lambda 200$–400 nm onto the upper layer samples (referred to as FS, "fully simulated"). Below, two identical sets of samples were kept in the dark. This middle and bottom layer experienced all simulation parameters except UVR-exposure (referred to as DS, "dark simulated"). While the BIOMEX samples (Fig. 1.4c, d) were glued to the EXPOSE-R2 compartment with a space-proofed two-component glue (Wacker® RTV-S 691 A/B, Wacker Chemie AG, Germany), the SVT samples were integrated without glue (Fig. 1.4a, b).

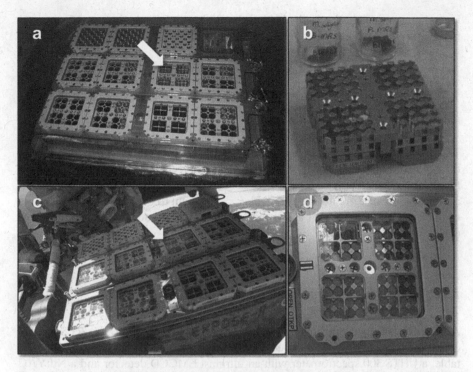

Fig. 1.4 Sample set-up for SVT and LEO-space exposure. (**a**) Sample arrangement of the biogenic compounds in tray 2 compartment 2 in the EXPOSE-R/E hardware at MUSC, DLR Cologne with tray 2, compartment 2 in the center (arrow) as used for SVT and MGR simulation procedures. (**b**) Three-layered carrier of the tray 2, compartment 2 after integration of the samples for SVT but before integration into the compartment. The MGR used the same configuration. (**c**) EXPOSE-R2 hardware outside the Zvezda module of the ISS with tray 2 compartment 2 in the center (arrow). (**d**) Close-up of tray 2, compartment 2 of EXPOSE-R2 with the P-MRS and S-MRS samples in the left and right lower half, respectively (**c**, **d** photographs taken at EVA #40 after removal of the EXPOSE-R2 protective cover at 22.10.2014 by Cosmonauts Alexandr Samokutyaev and Maksim Surayev)

1.6   Mission Ground Reference (MGR) Experiments

A simulation of the flight mission, the Mission Ground Reference (MGR), was performed at DLR's Microgravity User Support Center in Cologne (Rabbow et al. 2016). Duplicates of the flight samples were kept in flight-similar hardware, which was exposed to conditions mimicking the flight mission environment, based on available data obtained directly in space outside the ISS and within the limits of the simulation facilities. No temperature data was available for the dark outgassing period, as data acquisition systems were turned off to avoid heating of the samples under the cover. During the simulation of this period and until data was available, MGR trays were kept at temperatures of about 15 °C. Afterward, MGR conditions lagged approximately 2 months behind the flight mission conditions. More information on MGR parameters can be found in Rabbow et al. (2017).

1.7 Spectroscopic Analysis Before and After Long-Term Space Flight

The different spectroscopic analysis on the investigated samples were done mainly by Raman- and FTIR-spectroscopy but also analyzed by studying the sample's fluorescence behavior with specific CLSM techniques. The selected instruments are chosen, because in future they will play a significant role in the in situ exploration of planetary surfaces as well as for the search for life on other planets and moons (Table 1.4).

1.7.1 Raman Instrumentation

All Raman spectra were obtained with a confocal Witec alpha 300 RLS system consisting of a microscope equipped with a $10\times$ AFM objective, a piezo-driven scan table, a UHTS 300 spectrometer with an ultrafast EMCCD detector and a Nd:YAG laser. The excitation wavelength of the laser was 532 nm and the focus size at the sample was 1.5 µm. The excitation wavelength of the RLS and the laser power at the samples which was adjusted to 1 mW resembled the expected laser energy of the Pasteur Payload RLS of the ExoMars Rover (Rull et al. 2017; Vago et al. 2017). The Raman signal was collected in a spectral interval of 100–$3800\,\text{cm}^{-1}$ and with a spectral resolution of 4–$5\,\text{cm}^{-1}$. All measurements were obtained at ambient temperature and atmosphere but in different modes. Image scans of 100×100 µm with

Table 1.4 Scientific verification tests (SVT)

SVT	Duration	Pressure	Atmosphere	Temperature (T)	Irradiation
Tray 1	December 2013–January 2014, 38 day	Vacuum pressure at 4.1×10^{-5} Pas		T-cycles between -25°C (16 h in the dark) and +10 °C (8 h during irradiation)	The upper layers of each tray $UVR_{200\text{–}400nm}$ with $1271\,\text{W m}^{-2}$ ($5.7 \times 10^{5}\,\text{kJ m}^{-2}$) for 5924 min The lower layers of the trays were kept in dark
Tray 2			Mars atmosphere (95.55% CO_2, 2.7% N_2, 1.6% Ar, 0.15% O_2 and ~370 ppm H_2O at 1 kPa)	-23 °C	

104 single spectra measurements (10×, 0.1 s) as well as 200 μm line scans with 100 single measurements (1 s) were performed and analyzed with various filter scan ranges. Additional measurements were performed as time series and single spectra. In contrast to the EVT/SVT samples, the pure compounds were mostly measured with time series (10×, 50×, 100× for 1 s) with single spectra measurements (for 1, 2, or 5 s) and with variable laser energies 0.25–14 mW. Since single spectra and time series measurements are biased by the subjective choice of the measurement spot and thus may give low statistical significance, we choose line scans and, in case of β-carotene, imaging scans to multiply measurements and gain more reliable results of the SVT samples. For line scans, 5–10 scans in equal distance from the pellet margin were chosen randomly to obtain 500–1000 measurements. For 100 × 100 μm imaging scans, five areas for each pellet were randomly chosen, measured, and visualized by using five different filter ranges around the very strong and weak peaks of the investigated biomolecules (filter ranges from −15 to +15 cm^{-1} of each peak). These images were subdued to black/white binarization (threshold of 0.95, ImageJ software, W. Rasband, NIH, USA) to quantify the prevalence of the investigated biomolecule in terms of the pixel metering. The method was successfully cross-checked at a spectral range were no investigated biomolecule peak occurred revealing complementary images. The Raman spectra were baseline-corrected and annotated using ACD Labs and statistical analyses as two-sided student's t-test on significance with H0 > 0.05 were performed with GraphPad InStat 3 and Microsoft Excel.

1.7.2 Absorption Fourier Transform Infrared Spectroscopy (FTIR) and Attenuated Total Reflection Fourier Transform Infrared Spectroscopy (ATR-FTIR)

Besides Raman spectroscopy also other methods are well-known to be used for detection of biomolecules. For the detection of bacterial cellulose the FTIR spectroscopy was used to verify if such kind of technology could also be relevant for future space missions (de Vera 2020). The IR absorption analysis was carried out, using a Bruker 113v Fourier Transform IR spectrometer. The measurements were performed at room temperature in the range 4000–400 cm^{-1} with a spectral resolution of 1.0 cm^{-1} (Zaets et al. 2014). The absorption spectra of the samples obtained by ATR-FTIR enable to get information about the macromolecular composition. The ATR-FTIR method makes it possible to increase the analysis of the thickness of the investigated material due to the multiple reflectance of a beam from a prism surface. Dry samples of lab and space exposed samples were mounted on a KRS-5 prism. The prism with the sample was clamped in a special holder for ATR studies. The ATR-FTIR analyses were carried out like mentioned previously with a Bruker-113v Fourier Transform spectrometer. Measurements were performed at room temperature in the range of 500–4000 cm^{-1} with a spectral resolution of 1.0 cm^{-1}. The accuracy of determination of the line position was ± 1 cm^{-1} (Orlovska et al. 2020).

1.7.3 Confocal Laser Scanning Microscopy (CLSM): Fluorescence Analysis

The fluorescence analysis method, which was selected, was performed by a confocal laser scanning microscope coupled with a spectral analysis software (CLSM-λscan, see Baqué et al. 2014). This method enabled the observation of chlorophyll and its detectability also after UV irradiation. Small sample fragments (about 2 mm^2) were put onto slides and examined using a CLSM (Olympus Fluoview 1000 Confocal Laser Scanning System). Autofluorescence of photosynthetic pigments such as chlorophyll a, phycobiliproteins and also minerals was investigated by successively exciting the samples with 488-nm, 543-nm and 635-nm lasers, and collecting the emitted fluorescence in three channels: 503–524 nm, 555–609 nm, and 655–755 nm. Three-dimensional images were captured every 0.5 μm and processed with Imaris v. 6.1.0 software (Bitplane AG Zürich, Switzerland) to obtain maximum intensity projections. The spectral analysis of regions of interest (ROI) was performed using the 543-nm laser at 54 % of the maximum power (=0.54 mW) and collecting the emission from 553 to 800 nm, and mean fluorescence intensity (MFI) was measured. Curve plotting was performed using the GraphPad Prism program (GraphPad Software, San Diego, CA).

1.8 Results: Biomolecules as Biosignatures

Before presenting an overview on some results on investigated lab and/or space exposed biomolecules serving as potential biosignatures/bio-traces, it has to be emphasized that still at this stage of the finalization of this article analysis and interpretation is going on. In reference of progressing improvement of methodology and data analysis, a significant amount on images coupled with spectroscopic accumulative spectra data is under investigation. Discussions are running with questions how also to realize a significant nationally and internationally usable biosignature data base for improved support of future life detection missions in the solar system (see details also in de Vera et al. 2019a). Therefore, not all results obtained from the space exposed samples are published and available as reference for this summarizing article. But in the following chapters the main tendencies and still published data will give a promising view on the future use of space platforms such as EXPOSE on the ISS and others which may be placed beyond the Low Earth

Orbit with locations such as in orbit around the Moon or even on the surface of the Moon (de Vera et al. 2012).

1.8.1   Some Outcome From the Raman Spectroscopy Analysis on Methanogen Archaea, Their Proteins, Lipids And Their Calcite Production

The ExoMars mission will deploy a rover carrying a Raman spectrometer among the other analytical instruments in order to search for signatures of life and to investigate the Martian geochemistry. Raman spectroscopy is known as a powerful nondestructive optical technique for biosignature detection that requires only little sample preparation (Serrano et al. 2014).

According to the pre-flight results on the archaeon *Methanosarcina soligelidi* SMA-21 surprising results were achieved. These chosen organisms for use in space flight have shown on the single cell level to be in total detectable by Raman spectroscopy (Serrano et al. 2014, 2015). The results even have shown that we have to differentiate between different stages of the investigated cells in their life cycle which particularly means differences were observed according to their progressing growth phases (Fig. 1.5). During Methanosarcina's different growth phases the following different observations were made:

1. Early exponential phase: protein- and lipid-rich particles are detectable.
2. Late exponential phase: small increase of lipid-rich particles in the cell environment.
3. Stationary phase: a significant increase of lipid-rich particles and calcite.
4. Senescent phase: measured samples are dominated by calcite minerals.

We can conclude that the applied Raman technology is able to detect extant life, what should be also of significant relevance for the search for life on Mars and the Icy Moons of Jupiter and Saturn. Another outcome was that it is also possible to differentiate between different methanogen archaea species by the use of Raman spectroscopy (Serrano et al. 2015). The results on the space exposed samples are pending, because of the above mentioned reasons (see Sect. 1.8) and because of budgetary and personal lacks to finalize these time and resources consuming investigations. But according to the pre-flight results we could conclude that in reference to the more developed biosignature data base detailed protocols are needed to take into account the changing of the spectroscopic fingerprints of the investigated organisms during their life cycle, of their position as specie in the phylogenetic tree and the changings in the environmental conditions.

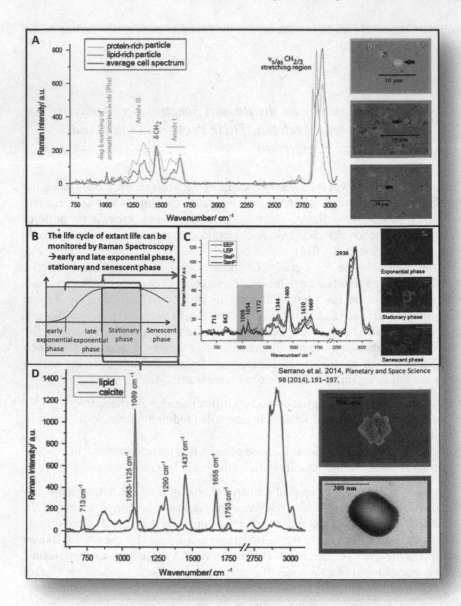

Fig. 1.5 Different appearance of the methanogen archaeon *Methanosarcina soligelidi* SMA 21 according to their life cycle; (**a**) Appearance of different spectra of the same specie according to the presence of proteins or enrichment of lipids. (**b–d**) detailed analysis of appearance and disappearance of specific spectra in reference to the growth phase stages (early exponential phase (EEP), late exponential phase (LEP), stationary phase (StaP), and senescent phase (SenP)). The enrichment in calcite is a hint to bio-fossilization of this specie (**d**); figures adapted from Serrano et al. (2014)

1.8.2   CLSM Fluorescence Analysis on Photosynthetic Pigments of Cyanobacteria

A reduction in the emission spectra of dried cells of the cyanobacterium *Chroococcidiopsis* sp. CCMEE 029 mixed with P-MRS or S-MRS was observed after exposure to a Mars-like UV flux (Billi et al. 2019). The reduction rate is slightly elevated in samples exposed to MGR, obviously due to the higher UV dose, which was simulated because of lack of some data sequences from space. Dried cells mixed with P-MRS or S-MRS and stored in the air-dried state under laboratory conditions for about 900 days showed a much more elevated reduction in the photosynthetic pigment fluorescence compared to samples exposed to Mars-like conditions in space and ground-based simulations in the bottom layer of the hardware (Fig. 1.4b). This could be explained as follows: cells accumulated oxidative damage during the prolonged dried storage under laboratory conditions in the oxygen rich surrounding atmosphere that did not occur under the Mars-like gas mixture consisting of 95.55% CO_2 (Billi et al. 2019). A reduction in the photosynthetic pigment fluorescence was also not observed in dried cells stored under laboratory conditions during the SVT (Baqué et al. 2016), may be also due to the shorter storage period (89 days) as compared to the EXPOSE-R2 mission (Fig. 1.6).

A comparison of the CLSM-λ-scan of the Martian regolith simulant/cell mixtures and their ground-based controls showed that in contrast to the S-MRS samples, dried cells that were mixed with P-MRS and exposed to Mars-like UV flux showed areas with an emission spectrum reduced about 35% of the original level observed for the control. Better protected areas were also observed on samples of dried cells mixed with P-MRS and exposed to the SVT. After this experiment the fluorescence was reduced only to 50% of that of hydrated cells; this finding was supposed to be due to the thinner grain size of P-MRS (compared to S-MRS), which might have led to thicker layers with enhanced shielding due to preparation of the cells/mineral mixture (Baqué et al. 2016). A reduction in the photosynthetic pigment autofluorescence of about 3.5% and 16.7% compared to the control was observed after the SVT experiment which has realized an exposure of the dried cells mixed with S-MRS or P-MRS to 570 MJ/m^2 of UV radiation (200–400 nm). The increased protective role of P-MRS might have been due to the fact that it contained a large fraction of clay minerals (notably montmorillonite, which accounts for 45% of the dry weight; see Table 1.2, Billi et al. 2019), which are considered to be good matrices for organic matter preservation in that they adsorb water molecules that might otherwise oxidize organic molecules and strongly absorb organic material (see, e.g. Hedges and Keil 1995; Farmer and Des Marais 1999).

According to these results obtained from lab and space research we can conclude, that also the protective or reactive/destructive role of the surrounding environmental properties has to be taken into account for correct data interpretation. This should guide us to select the correct landing sites on, e.g. Mars where promising mineral composition with protective properties for biomolecules could be favored to be explored for potentially preserved biosignatures. Also the fluorescence detection

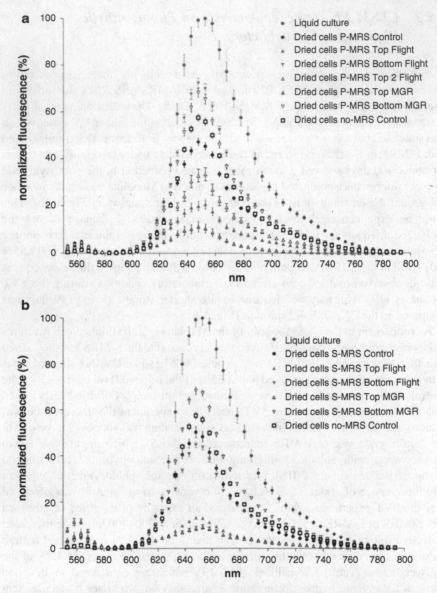

Fig. 1.6 Emission spectra of photosynthetic pigments in dried Chroococcidiopsis cells mixed with (**a**) P-MRS and (**b**) S-MRS or unmixed with minerals stored under laboratory conditions (P-MRS control, S-MRS control, and no-MRS control) were obtained after CLSM-λ-scan (instrument: 543 nm laser at 0.54 mW). Results shown here were obtained after exposure to Mars-like conditions in LEO (flight samples) or in MGR (mission ground reference). Data points represent normalized fluorescence intensity at 653 nm $\pm$ standard error for $n \geq 15$ cells as a function of emission wavelength (figure adapted from Billi et al. 2019, Astrobiology)

method is promising to be applied in future space missions with the main goal to search for life.

1.8.3   Cellulose Detection by ATR-FTIR after Exposure to Simulated Space/Mars-like Environment Realized in the Lab and in Space

According to the investigations performed within this study, cellulose was detectable by ATR-FTIR before and after exposure to the simulated planetary and space conditions in the lab and in space (personal communication and Orlovska et al. 2020 submitted). Even in contact to anorthosite, the cellulose contacting regolith is not much disturbing the observed spectra of cellulose. It could be clearly distinguished from the surrounding regolith (Zaets et al. 2014).

Previous investigations showed through spectral characterization that mineralized bacterial cellulose samples were changed diagenetically, but its feature characteristics could be recognized by FTIR spectroscopy (Zaets et al. 2014). A series of pre-flight experiments simulating the influence of space/Mars-like factors clearly showed just very small differences in the FTIR spectra between treated and laboratory specimens, which means that the stability of cellulose is very significant, because the ATR-FTIR method is also used for impurity investigations (Fuller et al. 2018).

1.8.4   Detection of Melanin by Raman Spectroscopy

Because results obtained from melanin samples which were exposed in space on the ISS were just submitted to a journal at the time the present article was written, an example could be mentioned here based on the results obtained through the EVT and SVT pre-flight tests. Those results have shown that irradiated melanin from fungal colonies of *Cryomyces antarcticus* under Mars-like conditions and embedded in the two Mars-Regolith-Simulants (P-MRS and S-MRS, see Table 1.2) were not affected by the environmental conditions (Pacelli et al. 2019, see Fig. 1.7). A personal communication (Pacelli et al. 2020, submitted) has confirmed also the same results in reference to the samples exposed on the ISS which were also additionally confirmed by other life detection methods such as IR-, Fourier Transform Infrared Resonance (FTIR) spectroscopy and by Gas Chromatography associated to Mass Spectrometry.

The use of different spectroscopy methods has even shown differences in the detection of molecule groups or derivates. In detail the following observations were made:

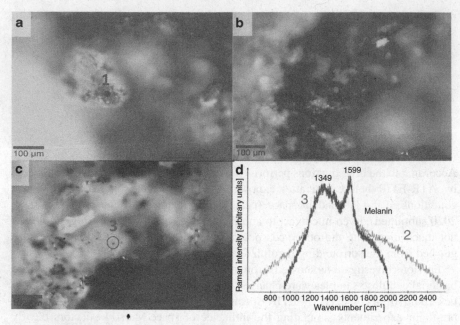

Fig. 1.7 Melanin Raman spectra (**d**, peaks 1349 cm^{-1} and 1499 cm^{-1}) obtained from *C. antarcticus* grown on OS, S-MRS, and P-MRS at maximum irradiation (respectively, **a**, **b**, and **c**) on selected point. Measurement time and number of spectra per point on sample were (10 s, 5×)

- IR spectroscopy showed the presence of carboxylic acids and amides, these molecules are not recorded in the Raman analysis.
- Raman analyses are dominated mainly by the melanin spectrum;
- The presence of melanin can also be found by FTIR spectroscopy, but it is difficult to unequivocally assign the bands to the presence of melanin due to vibration overlapping with other organic compounds and minerals.
- Identification of carboxylic acids by IR- and FTIR spectroscopy was confirmed by the GC-MS analyses indicating the presence of Eumelanin.

1.8.5   Detection of the Carotenoid Deinoxanthin by Raman Spectroscopy After Space/Mars-Like Exposure

Besides a number of studies on the stability and detectability of carotene after space exposure and during pre-flight investigations in the laboratory simulation experiments where we were able to show, that carotenes within cyanobacteria embedded in a Mars regolith mixture are stable and detectable by Raman spectroscopy (Baqué et al. 2020; Böttger et al. 2012, 2014), in this paragraph it should be highlighted the characteristics of a specific carotenoid, the Deinoxanthin within the extremely

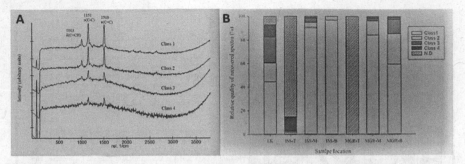

Fig. 1.8 Raman spectroscopic detection of Deinoxanthin after exposure on the ISS and mission ground control: (**a**) Categorization of spectra and classification in different classes. (**b**) Evaluation of the deinoxanthin signal intensity according to the defined classes following ground simulations and samples exposed to outer space conditions during the EXPOSE-R2 mission. Eight hundred and seventy five spectra were evaluated for each bar and scored according to the determined signal class. N.D. = no signal could be determined. Sample location indicates, whether samples were exposed to outer space (ISS) or part of the mission ground reference (MGR). LK = laboratory control. T, M, and B describe the sample position within the tray as follows: T = top; M = middle; B = bottom. Only top samples were exposed to solar radiation (source: Leuko et al. 2017)

radiation resistant, Mars-relevant and astrobiological model organism *Deinococcus radiodurans* after been exposed on the ISS.

It comes out that in this study the carotene derivate, the Deinoxanthin, was even detectable after exposure to space and Mars-like conditions realized directly in the Low Earth Orbit on the International Space Station (Leuko et al. 2017). Because of its stability if embedded in the cell matrix and its significant signal in the Raman spectra, this molecule could be classified as a real biosignature, where its signal could serve as reference in a biosignature database within the future space missions to Mars (see Fig. 1.8).

1.9 Conclusions

It has to be emphasized that the first attempt to select and investigate systematically biomolecules from their way out of their original habitats in their planetary analog field sites up to the planetary simulation laboratories and finally into space should be classified as a big success. It was shown that the preparation of astrobiological experiments in space is time consuming and energy/resources demanding. Intense pre-flight preparations are necessary. This includes the right selection of samples on planetary analog field sites mainly in remote places of the Arctic, Antarctic, and High altitude mountain regions, further selection through laboratory studies and pre-flight space simulation tests before sending the right and promising samples into space to perform following investigations after space exposure in Low Earth Orbit on the ISS. But if we have to look on the scientific outcome of such a mission, it has to be clarified that the performed investigations are absolutely not meaning that those

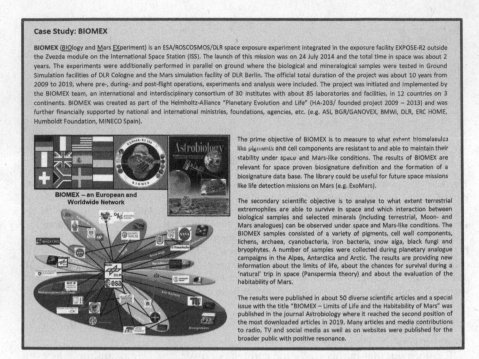

Case Study: BIOMEX

BIOMEX (BIOlogy and Mars EXperiment) is an ESA/ROSCOSMOS/DLR space exposure experiment integrated in the exposure facility EXPOSE-R2 outside the Zvezda module on the International Space Station (ISS). The launch of this mission was on 24 July 2014 and the total time in space was about 2 years. The experiments were additionally performed in parallel on ground where the biological and mineralogical samples were tested in Ground Simulation facilities of DLR Cologne and the Mars simulation facility of DLR Berlin. The official total duration of the project was about 10 years from 2009 to 2019, where pre-, during- and post-flight operations, experiments and analysis were included. The project was initiated and implemented by the BIOMEX team, an international and interdisciplinary consortium of 30 institutes with about 85 laboratories and facilities, in 12 countries on 3 continents. BIOMEX was created as part of the Heimholtz-Alliance "Planetary Evolution and Life" (HA-203/ founded project 2009 – 2013) and was further financially supported by national and international ministries, foundations, agencies, etc. (e.g. ASI, BGR/GANOVEX, BMWi, DLR, ERC HOME, Humboldt Foundation, MINECO Spain).

The prime objective of BIOMEX is to measure to what extent biomolecules like pigments and cell components are resistant to and able to maintain their stability under space and Mars-like conditions. The results of BIOMEX are relevant for space proven biosignature definition and the formation of a biosignature data base. The library could be useful for future space missions like life detection missions on Mars (e.g. ExoMars).

The secondary scientific objective is to analyse to what extent terrestrial extremophiles are able to survive in space and which interaction between biological samples and selected minerals (including terrestrial, Moon- and Mars analogues) can be observed under space and Mars-like conditions. The BIOMEX samples consisted of a variety of pigments, cell wall components, lichens, archaea, cyanobacteria, iron bacteria, snow alga, black fungi and bryophytes. A number of samples were collected during planetary analogue campaigns in the Alps, Antarctica and Arctic. The results are providing new information about the limits of life, about the chances for survival during a 'natural' trip in space (Panspermia theory) and about the evaluation of the habitability of Mars.

The results were published in about 50 diverse scientific articles and a special issue with the title "BIOMEX – Limits of Life and the Habitability of Mars" was published in the journal Astrobiology where it reached the second position of the most downloaded articles in 2019. Many articles and media contributions to radio, TV and social media as well as on websites were published for the broader public with positive resonance.

Fig. 1.9 BIOMEX as an example of a successful space experiment on the ISS

investigations are wasted energy and time. About 50 articles are published until know if only taken into account the space experiment BIOMEX on EXPOSE-R2/ISS, which served as example in this article (Fig. 1.9).

Besides of that also other experiments were done in parallel on EXPOSE-R2, as there were BOSS (P-I: DLR, Germany), PSS (P-I: LISA, France), Biodiversity (P-I: IMBP, Russia) with associated hardware experiments like PDP, BIOCHIP, DEPTH DOSE (all DLR, Germany), R3D-R2 (SRTI-BAS, Bulgaria), and PPOs (ESA/ESTEC, Netherlands; more details to all experiments: see Rabbow et al. 2017). Each of them has also a significant scientific and published outcome so that the knowledge gain should now be published largely in about more than 100 papers if only taken into account the mission EXPOSE-R2. By this it should be mentioned, that in the last two decades a number of other experiments were also done with similar exposure platforms on the ISS (EXPOSE-E: Rabbow et al. 2012 and EXPOSE-R1: Rabbow et al. 2015), where also a significant scientific knowledge on the previously investigated samples was gained. In the following paragraphs (Sects. 1.9.1, 1.9.2, and 1.9.3) we should have a closer look on the benefits we gained through the research done within the presented biosignature detection research of the BIOMEX experiment. This might serve as example that the work itself on such projects but also the results are significantly relevant for the broader society allowing scientific education and jobs as well as to gain new insights on the use of methods to be used applied also in other disciplines and the benefits of the results themselves.

Except these clear benefits, it should also not be a secret that still improvements of the existing hardware and instrumentation in space is needed. In reference to one of the main two goals of BIOMEX, as, e.g. the detection of biosignatures, it has also to be clarified that the EXPOSE-R2 and previous facilities on the ISS or even on satellites were not well equipped with the most relevant organic and biomolecule detection instruments for in situ measurements allowing a detection performance directly in space. It should be mentioned that there is a significant lack on data we should gain directly during the space exposure procedure. For a complete protocol supporting a well-established data base of biosignatures and traces which is containing finally the list of chosen instruments, data obtained by planetary analog field site characterization, data gained by planetary simulation experiments in the lab, and data obtained after space exposure we also need data obtained during the performed mission. No life detection instruments as presented in this article are until now operating in situ in space on the ISS during space exposure. So in case of some affecting events on the investigated biomolecules, we would not be able to know the exact point in time and the real final dose, where the biomolecules could have been damaged through the environmental conditions in space. We can just say, if the molecule was affected and damaged in a certain exposure time frame. So, there is a lot of space to improve the space experiments. A new generation of space investigation in LEO and beyond still has started with preliminary work to realize also in situ measurements on biological activity and detection of organics during space exposure (example OREOcube; Elsaesser et al. 2014). Further activities are also still preparing this new era of in situ operations on EXPOSE-like platforms as it is actually the case for the new platform EXPO on the Bartolomeo platform at the European Columbus module of the ISS. The Science Modules 1 for the experiments IceCold and ExoCube allow the in situ monitoring of active microbial cultures in the LEO radiation environments while exposed to extraterrestrial solar UV (Taubner et al. 2020).

1.9.1 Low Earth Orbit Experiments Supporting Future Search for Life on Mars

It is rarely well-known that performed astrobiological experiments in Low Earth Orbit have also direct relevance for further robotic space exploration missions in the Solar system. The broader public mainly knows about the studies on resistance of life and the questions on the potential to use tested organisms for biological life supporting systems. Its relevance for long-term human space flight missions is also well-known. Human space activities are mainly thought to be preparations of the astronauts for long term travels through space allowing them to do investigations and maintenance of the flight vehicle during longer stays in space as well as doing the necessary Extravehicular Activities (EVAs) for installing hardware outside the flight facilities. But besides of that, also the scientific operations and results are of high importance to improve exploring robotics and instruments for application on planetary bodies such as Mars or the icy moons of Jupiter, Saturn, Uranus or Neptune. In the vicinity of Earth, investigations are easy to operate and to control and all cases of a robotic mission could be simulated in micro-gravity under extreme radiation and vacuum conditions. Also instrument tests and the use of samples with relevance to planetary research are of high interest to the scientific community. Scientists need to know how the instruments are able to operate directly in space and how do behave samples under the space conditions. In orbit all space relevant environmental parameters are freely available and could be measured and observed simultaneously, whereat on ground it is not feasible at any existing laboratory to simulate this in total.

The idea to test biomolecules in LEO was a consequence of these considerations. If life arose or even could have originated once on Mars, extant life should leave or extinct life should have left traces of its existence on this planet. But how do we recognize this in an environment, so different from the terrestrial one. One approach is testing detection instruments, organisms, biomolecules, and organics under simulated conditions in specific planetary simulation laboratories. A much better approach is to test such instruments and samples directly in space. In Earth orbit it is still an approach where a much more important solar radiation income is affecting the tested samples compared to Mars or the outer solar system, but the influence of the galactic radiation is less important compared to the orbits further out in the solar system because of Earth's protecting magnetic field. Thus the correct calculation should give a very good approximation to the real picture we should get on the surface of Mars and might allow to get a feeling of the real final doses hitting the investigated samples, because the observed radiation dose is a matter of distance and time.

Based on the results, which were shown in the previous paragraphs of this article, we could say that surprisingly a significant number of selected biomolecules could serve as biosignature which is able to resist space and Mars-like conditions tested directly in space. The samples had stable detectable biomolecules, because the exploration mission relevant instruments were able to detect them also after the exposure under space and Mars-like conditions on the ISS.

1.9.2   Low Earth Orbit Experiments Supporting Future Search for Life on Icy Ocean Worlds

After these successful biosignature investigations on EXOPOSE-R2 within the BIOMEX project, it was clear to study now further fossils with Mars-relevance and biomolecules and organics which should be of interest to look for on the icy ocean worlds of the outer solar system.

Fig. 1.10 Logo of the new
Exposure Experiment
BioSigN on EXPO/ISS
(launch foreseen 2024)

In the case of Saturnian moon Enceladus, the analysis by the Cassini mission of the South Polar Plume has led to the detection of complex organics (Waite Jr et al. 2009). For Europa, it has been suggested that the remote detection of biosignatures might be possible through spectroscopy of ocean material that reached the surface through fractures or via briny diapirs.

Such extraterrestrial habitats, which might harbor life forms or could obtain traces of ancient life as to be supposed on the planet Mars or any niches resembling to, e.g. cracks and fissures in the outermost crust layers of the icy moons and which store possible ingredients or deposits of the geysers and fountains of the icy moons (McKay et al. 2014), must be detectable in future space exploration missions (de Vera et al. 2019a).

Based on this knowledge, a new space exposure experiment named BioSigN was proposed for the new EXPO facility on the ISS and approved by ESA, where a selection of new samples with also Icy Moon relevance was suggested and also performed. The preparation for this experiment with a supposed launch in 2024 has currently started (Fig. 1.10). Studies on new biomolecules but also on organics with potential relevance for the origin of life are included in the chosen BioSigN molecule set.

1.9.3   Earth Benefit of Astrobiological Space Research

The new generation of space exposure experiments on the ISS has a significant impact on different scientific disciplines and also on education, the career of young scientists and a job machine. Further, the public interest increased.

In reference to the scientific value, the following can be emphasized:

1. Cooperation between scientist in the domains of Polar Research, Ocean Research, and Space Research were achieved, leading to new multi functional instruments for exploration of a diversity of extreme environments
2. Positive evaluation of space relevance of life detection instruments for exploration missions with the main focus to search for life in the universe
3. Determination of the right way for usable biosignature and bio-trace data bases could be elaborated
4. Significant valuable knowledge gain on the stability of biomolecules and their potential protective role against harsh environmental conditions (such as desiccation and high irradiation dose)/relevance for cosmetic industry might be a consequence
5. Results on investigations on probiotic cultures under extreme conditions of space indicates that these types of potential nutrient ingredients are stabile during long duration space missions and could trigger and stabilize the strength of the immune system of astronauts during space travels
6. Results were relevant also to cooperating other projects and vice versa (examples: STARLIFE, Helmholtz Alliances Planetary Evolution and Life/ROBEX, ERC HOME, etc.

In reference to the relevance for education, the space exposure experiment BIOMEX can also deliver the following highlights (see also Table 1.5), which are showing that about in total 34 students/young researcher were trained and employed within this project:

Table 1.5 Education/training/degrees during BIOMEX

Countries	Master students	PhD students	PostDocs	Totally trained
France	1			1
Germany	2	7	5	14
Italy	2	4	2[a]	8
Netherlands			1[a]	1
Spain		2		2
Sweden	1			1
Switzerland		1		1
United Kingdom		4		4
USA		2[b]		2
In total	6	20	8	34

[a]Partly also in Germany
[b]Partly also in Italy

Table 1.6 Number of awards during the BIOMEX project

Countries	Research/oral presentation	Poster	In total (awarding org.)
Germany	1	2	3 (EANA)
Italy	2		2 (EANA + SCAR)
Spain	1		1 (EANA)
In total	4	2	6

1. 6 Master students were trained and did their investigations within the BIOMEX project and obtained successfully their Master degree.
2. 20 PhD students were trained and work on the sub-projects of BIOMEX in their P-I/Co-I institutions; 12 finished their PhD directly in reference to the BIOMEX project and 8 were or are still involved in connected topic work and finalizing actually their PhD degree
3. 8 PostDocs were employed within BIOMEX (some also resulting from the big PhD pool mentioned in (2)).

The German group had the majority of young scientist, because it was the Principal Investigator nation with also the biggest group. It has to be mentioned that although Germany had the biggest group of trained students and young researcher, it was less supported by founding organizations which are directly involved in space research. Just 3 of the 14 young researchers were supported annually by space-related agencies and the rest had to get support through other financial resources such as by the support of scientific foundations or integrated in other projects. Italy and Spain were covered practically completely by space research supporting organizations. The other participating nations had the same split of founding support like Germany.

It has to be emphasized, that 6 of the above mentioned students and young researchers got an award for their internationally recognized research (see Table 1.6). Four of them got a research/oral presentation award (first and second price) and two of them a poster award at the EANA (European Astrobiology Network Association 2011–2017) and SCAR Biology 2017 (Scientific Committee of Antarctic Research). Details are mentioned in Table 1.6.

It has to be mentioned that some of the master and PhD students got also research travel awards through the ERASMUS+ program from the EU or a PostDoc position through the Humboldt-Stiftung.

This is an excellent achievement and it has to be emphasized that the majority of the young scientists also got follow-on jobs partly in the space science / research domain after achieved their degrees and awards.

General benefits for the investigators of BIOMEX:

1. Extended international network was created during BIOMEX (30 institutes from 12 nations on 3 continents, Fig. 1.11)
2. An increased pool of investigators (about 70 people) was created which allows the cooperation in building new project groups and new research projects

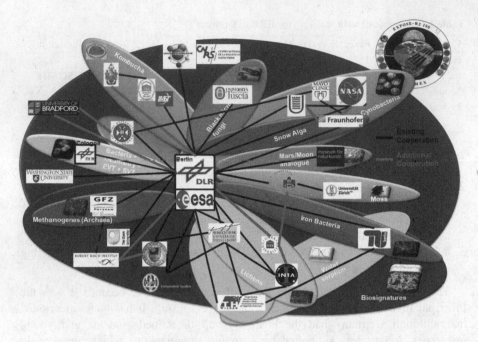

Fig. 1.11 BIOMEX—an European and Worldwide Network: black connecting balk shows existing cooperations between national and international institutes starting at the beginning of BIOMEX and red balks are indicating new formed cooperations during the BIOMEX project. Also new institutes joined during the project

3. Exchange of knowledge beyond the own expertise and the own scientific discipline
4. Interactions with astronauts were realized
5. High visibility through national and international outreach activities
6. Sharing of logistics, laboratories, facilities, instruments, and methods as well as samples for further investigations which originally were not planned and enlarged finally the amount of obtained results (Fig. 1.11)
7. BIOMEX meetings with their social events strengthened the identity of the international BIOMEX investigator community and their direct interactions
8. High number of scientific articles (about 50 articles in international scientific journals and books) were published.

References

Allen CC, Albert FG, Chafetz HS, Combie J, Graham CR, Kieft TL, Kivett SJ, McKay DS, Steele A, Taunton AE, Taylor MR, Thomas-Keprta KL, Westall F (2000) Microscopic physical biomarkers in carbonate hot springs: implications in the search for life on Mars. Icarus 147:49–67

Banfield JF, Moreau JW, Chan CS, Welch SA, Little B (2004) Mineralogical biosignatures and the search for life on Mars. Astrobiology 1(4):447–465

Baqué M, Verseux C, Rabbow E, de Vera JPP, Billi D (2014) Detection of macromolecules in desert cyanobacteria mixed with a lunar mineral analogue after space simulations. Orig Life Evol Biosph 44(3):209–221

Baqué M, Verseux C, Böttger U, Rabbow E, de Vera J-PP, Billi D (2015) Biosignature preservation of cyanobacteria mixed with phyllosilicatic and sulfatic Martian regoliths under simulated Martian atmosphere and UV flux. Orig Life Evol Biosph 46(2):289–310. https://doi.org/10.1007/s11084-015-9467-9

Baqué M, Verseux C, Böttger U, Rabbow E, de Vera J-P, Billi D (2016) Preservation of biomarkers from cyanobacteria mixed with Mars-like regolith under simulated Martian atmosphere and UV flux. Orig Life Evol Biosph 46:289–310

Baqué M, Hanke F, Böttger U, Leya T, Moeller R, de Vera J-P (2018) Protection of cyanobacterial carotenoids' Raman signatures by Martian mineral analogues after high-dose gamma irradiation. J Raman Spectrosc. https://doi.org/10.1002/jrs.5449

Baqué M, Napoli A, Fagliarone C, Moeller R, de Vera J-P, Billi D (2020) Carotenoid Raman Signatures are better preserved in dried cells of the desert cyanobacterium Chroococcidiopsis than in hydrated counterparts after High-Dose Gamma Irradiation. Life 10:83. https://doi.org/10.3390/life10060083

Beegle L et al (2015) SHERLOC: scanning habitable environments with Raman & luminescence for organics & chemicals. In: 2015 IEEE Aerospace Conference, Big Sky, MT, 2015, pp 1–11. https://doi.org/10.1109/AERO.2015.7119105

Billi D, Verseux C, Fagliarone C, Napoli A, Baqué M, de Vera J-P (2019) A desert cyanobacterium under simulated Mars-like conditions in Low Earth Orbit: implications for the habitability of Mars. Astrobiology 19(2):158–169

Boston PJ, Spilde MN, Northup DE, Melim LA, Soroka DS, Kleina LG, Lavoie KH, Hose LD, Mallory LM, Dahm CN, Crossey LJ, Schelble RT (2004) Cave biosignature suites: microbes, minerals, and mars. Astrobiology 1(1):25–55

Böttger U, de Vera J-P, Fritz J, Weber I, Hübers H-W, Schulze-Makuch D (2012) Optimizing the detection of carotene in cyanobacteria in a Martian regolith analogue with a Raman spectrometer for the ExoMars mission. Planet Space Sci 60:356–362

Böttger U, de la Torre R, Frias J-M, Rull F, Meessen J, Sánchez Íñigo FJ, Hübers H-W, de Vera JP (2014) Raman spectroscopic analysis of the oxalate producing extremophile Circinaria Gyrosa. Int J Astrobiol 13(1):19–27

Brasier M, Green O, Jephcoat A, Kleppe A, van Kranendonk M, Lindsay J, Steel A, Grassineau N (2002) Questioning the evidence for earth's oldest fossils. Nature 416:76–81

Cady SL, Farmer JD, Grotzinger JP, Schopf JW, Steele A (2003) Morphological biosignatures and the search for life on mars. Astrobiology 3(2):351–368

Cockell CS (2002) The ultraviolet radiation environment of Earth and Mars: past and present. In: Astrobiology. Springer, New York, pp 219–232

Cottin H, Kotler JM, Billi D, Cockell C, Demets R, Ehrenfreund P, Elsaesser A, d'Hendecourt L, van Loon JJWA, Martins Z, Onofri S, Quinn RC, Rabbow E, Rettberg P, Ricco AJ, Slenzka K, de la Torre R, de Vera J-P, Westall F, Carrasco N, Fresneau A, Kawaguchi Y, Kebukawa Y, Nguyen D, Poch O, Saiagh K, Stalport F, Yamagishi A, Yano H, Klamm BA (2017) Space as a tool for astrobiology: review and recommendations for experimentations in earth orbit and beyond. Space Sci Rev. https://doi.org/10.1007/s11214-017-0365-5

Dachwald B, Ulamec S, Postberg F, Sohl F, de Vera J-P, Waldmann C, Lorenz RD, Zacny KA, Hellard H, Biele J, Rettberg P (2020) Key technologies and instrumentation for subsurface exploration of ocean worlds. Space Sci Rev 216:83. https://doi.org/10.1007/s11214-020-00707-5

Dadachova E, Bryan RA, Huang X, Moadel T, Schweitzer AD, Aisen P, Nosanchuk JD, Casadevall A (2007) Ionizing radiation changes the electronic properties of melanin and enhances the growth of melanized fungi. PLoS One 2:e457 de Vera JPP (2020) The relevance of ecophysiology in astrobiology and planetary research: implications for the characterization of the habitability of planets and biosignatures. Habilitationsschrift. Uni Potsdam, Potsdam

de Vera J-P, Boettger U, de la Torre Noetzel R, Sánchez FJ, Grunow D, Schmitz N, Lange C, Hübers H-W, Billi D, Baqué M, Rettberg P, Rabbow E, Reitz G, Berger T, Möller R, Bohmeier M, Horneck G, Westall F, Jänchen J, Fritz F, Meyer C, Onofri S, Selbmann L, Zucconi L, Kozyrovska N, Leya T, Foing B, Demets R, Cockell CS, Bryce C, Wagner D, Serrano P, Edwards HGM, Joshi J, Huwe B, Ehrenfreund P, Elsaesser A, Ott S, Meessen J, Feyh N, Szewzyk U, Jaumann R, Spohn T (2012) Supporting Mars exploration: BIOMEX in low earth orbit and further astrobiological studies on the Moon using Raman and PanCam technology. Planet Space Sci 74(1):103–110

de Vera J-P, Schulze-Makuch D, Khan A, Lorek A, Koncz A, Möhlmann D, Spohn T (2014) Adaptation of an Antarctic lichen to Martian niche conditions can occur within 34 days. Planet Space Sci 98:182–190

de Vera JP, Baqué M, Billi D, Böttger U, Cockell CS, de la Torre R, Foing BH, Hanke F, Leuko S, Martinez-Frías J, Moeller R, Olsson-Francis K, Onofri S, Rettberg P, Schröder S, Schulze-Makuch D, Selbmann L, Wagner D, Zucconi L (2019a) A systematic way to life detection – combining field, lab and space research in low Earth orbit. In: Cavalazzi B, Westall F (eds) Biosignatures for astrobiology. Springer, Cham, pp 111–122

de Vera JP, Alawi M, Backhaus T, Baqué M, Billi D, Böttger U, Berger T, Bohmeier M, Cockell C, Demets R, de la Torre Noetzel R, Edwards H, Elsaesser A, Fagliarone C, Fiedler A, Foing B, Foucher F, Fritz J, Hanke F, Herzog T, Horneck G, Hübers H-W, Huwe B, Joshi J, Kozyrovska N, Kruchten M, Lasch P, Lee N, Leuko S, Leya T, Lorek A, Martínez-Frías J, Meessen J, Moritz S, Moeller R, Olsson-Francis K, Onofri S, Ott S, Pacelli C, Podolich O, Rabbow E, Reitz G, Rettberg P, Reva O, Rothschild L, Garcia Sancho L, Schulze-Makuch D, Selbmann L, Serrano P, Szewzyk U, Verseux C, Wadsworth J, Wagner D, Westall F, Wolter D, Zucconi L (2019b) Limits of life and the habitability of mars: the ESA space experiment BIOMEX on the ISS. Astrobiology 19(2):145–157

Edwards HGM, Hutchinson I, Ingley R (2012) The ExoMars Raman spectrometer and the identification of biogeological spectroscopic signatures using a flight-like prototype. Anal Bioanal Chem 404(6):1723–1731

Edwards HGM, Hutchinson IB, Ingley R, Parnell J, Vítek P, Jehlička J (2013) Raman spectroscopic analysis of geological and biogeological specimens of relevance to the ExoMars mission. Astrobiology 13(6):543–549. https://doi.org/10.1089/ast.2012.0872

Elsaesser A, Quinn RC, Ehrenfreund P, Mattioda AL, Ricco AJ, Alonzo J, Breitenbach A, Chan YK, Fresneau A, Salama F, Santos O (2014) Organics Exposure in Orbit (OREOcube): a next-generation space exposure platform. Langmuir 30(44):13217–13227. https://doi.org/10.1021/la501203g

Farmer JD, Des Marais DJ (1999) Exploring for a record of ancient Martian life. J Geophys Res Planets 104:26977–26995

Formisano V, Atreya S, Encrenaz T, Ignatiev N, Giuranna M (2004) Detection of methane in the atmosphere of mars. Science 306(5702):1758–1761

Fuller ME, Andaya C, McClay K (2018) Evaluation of ATR-FTIR for analysis of bacterial cellulose impurities. J Microbiol Methods 144:145–151. https://doi.org/10.1016/j.mimet.2017.10.017

Gadd GM (2007) Geomycology: biogeochemical transformations of rocks, minerals, metals and radionuclides by fungi, bioweathering and bioremediation. Mycol Res 111(1):3–49

Gessler NN, Egorova AS, Belozerskaya TA (2014) Melanin pigments of fungi under extreme environmental conditions. Appl Biochem Microbiol 50:105–113

Grenville-Briggs LJ, Anderson VL, Fugelstad J, Avrova AO, Bouzenzana J, Williams A et al (2008) Cellulose synthesis in Phytophthora infestans is required for normal appressorium formation and successful infection of potato. Plant Cell 20(3):720–738

Hauber E, Sassenroth C, de Vera J-P, Schmitz N, Jaumann R, Reiss D, Hiesinger H & Johnsson A (2019) Debris flows and water tracks in northern Victoria Land, continental East Antarctica: a new terrestrial analogue site for gullies and recurrent slope lineae on Mars. In: Conway SJ, Carrivick JL, Carling PA, de Haas T & Harrison TN Martian gullies and their earth analogues, Geological Society, London, 467(1):267

Hedges JI, Keil RG (1995) Sedimentary organic matter preservation: an assessment and speculative synthesis. Mar Chem 49:81–115

Hoefs J (1969) Natural calcium oxalate with heavy carbon. Nature 223:396

Kawano Y, Saotome T, Ochiai Y, Katayama M, Narikawa R, Ikeuchi M (2011) Cellulose accumulation and a cellulose synthase gene are responsible for cell aggregation in the cyano-bacterium Thermosynechococcus vulcanus RKN. Plant Cell Physiol 52(6):957–966

Lefèvre F, Forget F (2009) Observed variations of methane on Mars unexplained by known atmospheric chemistry and physics. Nature 460:720–723

Leuko S, Bohmeier M, Hanke F, Böttger U, Rabbow E, Parpart A, Rettberg P, de Vera J-PP (2017) On the stability of deinoxanthin exposed to mars conditions during a long-term space mission and implications for biomarker detection on other planets. Front Microbiol 8:1680

Marshall C, Emry J, Marshall A (2011) Haematite pseudomicrofossils present in the 3.5-billion-year-old apex chert. Nat Geosci 4:240–243

McKay CP, Anbar AD, Porco C, Tsou P (2014) Follow the plume: the habitability of Enceladus. Astrobiology 14(4):352–355

Meredith P, Sarna T (2006) The physical and chemical properties of eumelanin. Pigment Cell Res 19:572–594

Mumma MJ, Villanueva GL, Novak RE, Hewagama T, Bonev BP, DiSanti MA, Mandell AM, Smith MD (2009) Strong release of methane on Mars in northern summer 2003. Science 323 (5917):1041–1045

Navarrete JU, Cappelle IJ, Schnittker K, Borrok DM (2013) Bioleaching of ilmenite and basalt in the presence of iron-oxidizing and iron-scavenging bacteria. Int J Astrobiol 12(2):123–134

Nobles DR, Brown RM (2004) The pivotal role of cyanobacteria in the evolution of cellulose synthases and cellulose synthase-like proteins. Cellulose 11(3):437–448

Nobles DR, Romanovicz DK, Brown RM (2001) Cellulose in cyanobacteria. Origin of vascular plant cellulose synthase? Plant Physiol 127(2):529–542

Okuda K, Sekida S, Yoshinaga S, Suetomo Y (2004) Cellulosesynthesizing complexes in some chromophyte algae. Cellulose 11(3):365–376

Orlovska I, Podolich O, Kukharenko O, Zaets I, Reva O, Khirunenko L, Zmejkoski D, Rogalsky S, Barh D, Tiwari S, Góes-Neto A, Azevedo V, Brenig B, Ghosh P, de Vera J-P, Kozyrovska N (2020) Maintained robustness of bacterial cellulose but decline of its synthesis after exposure to Mars-like environment simulated outside the International Space Station. Astrobiology (in review)

Ourisson G, Nakatani Y (1994) The terpenoid theory of the origin of cellular life: the evolution of terpenoids to cholesterol. Chem Biol 1:11–23

Pacelli C, Selbmann L, Zucconi L, Coleine C, de Vera J-P, Rabbow E, Böttger U, Dadachova E, Onofri S (2019) Responses of the black fungus Cryomyces antarcticus to simulated mars and space conditions on rock analogues. Astrobiology 19(1):209–220. https://doi.org/10.1089/ast.2016.1631

Pacelli C, Cassaro A, Baqué M, Selbmann L, Zucconi L, Maturilli A, Botta L, Saladino R, Böttger U, Demets R, Rabbow E, de Vera, J-P, Onofri S (2020) Fungal biomarkers are detectable in Martian rock-analogues after space exposure: implications for the search of life on Mars. Nat Astron (in review)

Rabbow E, Rettberg P, Barczyk S, Bohmeier M, Parpart A, Panitz C, Horneck G, von Heise-Rotenburg R, Hoppenbrouwers T, Willnecker R, Baglioni P, Demets R, Dettmann J, Reitz G (2012) EXPOSE-E: an ESA astrobiology mission 1.5 years in space. Astrobiology 12 (5):374–386

Rabbow E, Rettberg P, Barczyk S, Bohmeier M, Parpart A, Panitz C, Horneck G, Burfeindt J, Molter F, Jaramillo E, Pereira C, Weiß P, Willnecker R, Demets R, Dettmann J, Reitz G (2015) The astrobiological mission EXPOSE-R on board of the International Space Station. Int J Astrobiol 14(1):3–16. https://doi.org/10.1017/S1473550414000202

Rabbow E, Parpart A, Reitz G (2016) The planetary and space simulation facilities at DLR Cologne. Microgravity Sci Technol 28:215–229

Rabbow E, Rettberg P, Parpart A, Panitz C, Schulte W, Molter F, Jaramillo R, Demets R, Weiß P, Wilnecker R (2017) EXPOSE-R2: the astrobiological ESA mission on board of the International Space Station. Front Microbiol 8:1533. https://doi.org/10.3389/fmicb.2017.01533

Rettberg P, Rabbow E, Panitz C, Horneck G (2004) Biological space experiments for the simulation of Martian conditions: UV radiation and Martian soil analogues. Adv Space Rev 33:1294–1301

Ross P, Mayer R, Benziman M (1991) Cellulose biosynthesis and function in bacteria. Microbiol Rev 55(1):35–58

Rothschild LJ (1990) Earth analogs for Martian life. Microbes in evaporites, a new model system for life on Mars. Icarus 88:246–260

Rull F, Maurice S, Hutchinson I, Moral A, Perez C, Diaz C, Colombo M, Belenguer T, Lopez-Reyes G, Sansano A, Forni O, Parot Y, Striebig N, Woodward S, Howe C, Tarcea N, Rodriguez P, Seoane L, Santiago A, Rodriguez-Prieto JA, Medina J, Gallego P, Canchal R, Santamaría P, Ramos G, Vago JL, RLS Team (2017) The Raman laser spectrometer for the ExoMars rover mission to mars. Astrobiology 17(6):627–654

Schopf J (1993) Microfossils of the Early Archean Apex chert: new evidence of the antiquity of life. Science 260:640–646

Schulze-Makuch D, Wagner D, Kounaves SP, Mangelsdorf K, Devine KG, de Vera J-P, Schmitt-Kopplin P, Grossart H-P, Parro V, Kaupenjohann M, Galy A, Schneider B, Airo A, Frösler J, Davila AF, Arens FL, Cáceres L, Cornejo FS, Carrizo D, Dartnell L, DiRuggiero J, Flury M, Ganzert L, Gessner MO, Grathwohl P, Guan L, Heinz J, Hess M, Keppler F, Maus D, McKay CP, Meckenstock RU, Montgomery W, Oberlin EA, Probst AJ, Sáenz JS, Sattler T, Schirmack J, Sephton MA, Schloter M, Uhl J, Valenzuela B, Vestergaard G, Wörmer L, Zamorano P (2018) Transitory microbial habitat in the hyperarid Atacama Desert. PNAS 115 (11):2670–2675

Serrano P, Hermelink A, Boettger U, de Vera J-P, Wagner D (2014) Single-cell analysis of the methanogenic archaeon Methanosarcina soligelidi from Siberian permafrost by means of confocal Raman microspectrocopy for astrobiological research. Planet Space Sci 98:191–197

Serrano P, Hermelink A, Lasch P, de Vera J-P, König N, Burckhardt O, Wagner D (2015) Confocal Raman microspectroscopy reveals a convergence of the chemical composition in methanogenic archaea from a Siberian permafrost-affected soil. FEMS Microbiol Ecol 91:126

Sohrabi M (2012) Taxonomy and phylogeny of the manna lichens and allied species (Megasporaceae). PhD thesis, publications in botany from the University of Helsinki

Stackebrandt E (2004) The phylogeny and classification of anaerobic bacteria horizon. Bioscience 2004:1–26

Stahl W, Sies H (2003) Antioxidant activity of carotenoids. Mol Asp Med 24(6):345–351

Taubner R-S, Olsson-Francis K, Vance SD, Ramkissoon NK, Postberg F, de Vera J-P, Antunes A, Camprubi Casas E, Sekine Y, Noack L, Barge L, Goodman J, Jebbar M, Journaux B, Karatekin Ö, Klenner F, Rabbow E, Rettberg P, Rückriemen-Bez T, Saur J, Shibuya T, Soderlund KM (2020) Experimental and simulation efforts in the astrobiological exploration of exooceans. Space Sci Rev 216:9. https://doi.org/10.1007/s11214-020-0635-5

Vago JL, Westall F, Pasteur Instrument Teams, Landing Site Selection Working Group, and Other Contributors (2017) Habitability on early mars and the search for biosignatures with the ExoMars rover. Astrobiology 17(6):471–510

Vassilev S, Vassilev C (1996) Occurrence, abundance and origin of minerals in coals and coal ashes. Fuel Process Technol 48:85–106

Wacey D, Kilburn M, Saunders M, Cliff J, Brasier M (2011) Microfossils of sulphur-metabolizing cells in 3.4-billion year old rocks of Western Australia. Nat Geosci 4:698–702

Waite JH Jr, Lewis WS, Magee BA, Lunine JI, McKinnon WB, Glein CR, Mousis O, Young DT, Brockwell T, Westlake J, Nguyen MJ, Teolis BD, Niemann HB, McNutt RL Jr, Perry M, Ip WH (2009) Liquid water on Enceladus from observations of ammonia and 40Ar in the plume. Nature 460:487–490

Ward C (2002) Analysis and significance of mineral matter in coal seams. Int J Coal Geol 50:135–168

Weber KA, Spanbauer TL, Wacey D, Kilburn MR, Loope DB, Kettler RM (2012) Biosignatures link microorganisms to iron mineralization in a paleoaquifer. Geology 40(8):747–750

Webster CR, Mahaffy PR, Atreya SK, Flesch GJ, Mischna MA, Meslin P-Y, Farley KA, Conrad PG, Christensen LE, Pavlov AA, Martín-Torres J, Zorzano M-P, McConnochie TH, Owen T, Eigenbrode JL, Glavin DP, Steele A, Malespin CA, Archer PD Jr, Sutter B, Coll P, Freissinet C, McKay CP, Moores JE, Schwenzer SP, Bridges JC, Navarro-Gonzalez R, Gellert R, Lemmon MT, MSL Science Team (2015) Mars methane detection and variability at Gale crater. Science 237(6220):415–417

Westall F (1999) The nature of fossil bacteria: a guide to the search for extraterrestrial life. J Geophys Res Planets 104:16437–16442

Westall F, Steele A, Toporski J, Walsh M, Allen C, Guidry S, McKay CD, Gibson E, Chafetz H (2000) Polymeric substances and biofilms as biomarkers in terrestrial materials: implications for extraterrestrial samples. J Geophys Res 105(10):24511–24528

Zaets I, Podolich O, Kukharenko O, Reshetnyak G, Shpylova S, Sosnin M, Khirunenko L, Kozyrovska N, de Vera J-P (2014) Bacterial cellulose may provide the microbial-life biosignature in the rock records. Adv Space Res 53:828–835

Zahnle K, Freedman RS, Catling DC (2011) Is there methane on Mars? Icarus 212:493–503

Zák K, Skála R (1993) Carbon isotopic composition of whewellite ($CaC_2O_4 \cdot H_2O$) from different geological environments and its significance. Chem Geol 106:123–131

Chapter 2
Habitability Tests in Low Earth Orbit

Abstract The science case for Mars in Planetary Research is mainly driven by the following questions: (1) is Mars habitable? (2) Could we be able to detect extant and/or extinct life on Mars? (3) Do biomolecules exist to be relevant biosignatures to look for on Mars? (4) Is the interplanetary transfer of life between the Earth-Mars-System possible (Lithopanspermia)? (5) Does the origin of methane on Mars come from potential organisms on this planet? (6) What are the interactions between radiation and water on the Martian surface? (7) How does the atmospheric photo-chemistry and the observed weather influence the Martian climate over short and longer time scales? Experiments performed during the EXPOSE-R2 mission particularly in the BIOMEX project tried to give answers mainly to Mars' habitability and life detection questions. In this chapter the main focus is on performed Low Earth Orbit (LEO) studies on the habitability of Mars and also consequences on potential interplanetary transfer of life within the Earth-Mars-System as well as to the limits of terrestrial life. The experiments compared to previously performed experiments on the ISS and satellites have shown differences on the organisms' response and allowed a differentiation on the evaluation of the strength and resistance of the tested biological samples. It comes out that depending on the species used differences in the behavior were obvious, even if they are classified to be of the same phylogenetic order, family or genus and even if they are living in planetary analog field sites. There could not be a generalized conclusion that specific terrestrial organism groups are more resistant and potentially suitable to live on Mars than others. If the final conclusion would be expressed like this: Mars seems to be habitable for methanogen archaea, bacteria, cyanobacteria, micro-fungi, and partly some lichens, we should be aware that this is just valid for some specific species and not for all species of those domains and their original planetary-relevant habitats. It also means that a more detailed analysis is needed in future. But based on the obtained results, also interesting application opportunities of some resistant organisms or even biofilms could play a role in triggering the immune system and serving as food for astronauts during long manned space missions as well as to be used for life supporting systems. Astrobiology is therefore directly also influencing the disciplines of terrestrial and potentially extraterrestrial agricultures, Life Sciences, and Space Medicine. An interaction through all space-related biology fields is highly appreciated for the gain of maximum benefits for space research, for our planet Earth and humankind.

J.-P. de Vera, *Astrobiology on the International Space Station*, SpringerBriefs in
Space Life Sciences, https://doi.org/10.1007/978-3-030-61691-5_2

Keywords Archaea · Bacteria · Eukaryotes · Habitability · Lithopanspermia · ISS · Earth-Mars-System · Mars · Low Earth Orbit (LEO)

2.1   Introduction

The habitability of Mars and other planetary bodies such as the icy moons is often evaluated and classified by disciplines of physicists, chemists, geologists, but rarely by biologists, although they should have the knowledge on the reference organisms, which could live in planetary environments mainly described by the mentioned scientific disciplines. If there is a general discourse on habitability, the broader scientific community should take into account what they are talking about. Habitability of a planet means from a biological point of view that some kinds of life forms are able to live in the described environment. But because not all kind of life forms are able to live for, e.g. on Mars, the biologists have to correct and to more differentiate at what degree a planet is habitable and for what kind of organisms this statement should be valid.

Based on this knowledge, experiments with microorganisms were started in different laboratories and facilities to approach as much as possible the planetary conditions of a planet of interest. In the last decades, the most familiar and easily realized simulation was done in reference to planet Mars (Jensen et al. 2008; Robledo-Martinez et al. 2012; de Vera et al. 2014a, b; besides others). This was a good basis to get first best approach on evaluating terrestrial organisms on their ability to resist, adapt, and live under Martian conditions. If such positive observations were done, Mars could from a biological point of view be classified as really habitable in reference to the investigated organisms.

But, how are the conditions, life have to face on Mars? We have to take into consideration that Mars has a very thin atmosphere with a pressure of about 6–8 mbar approaching vacuum conditions. Besides the missing thick atmosphere, there is no shielding by a global magnetic field what means that nearly complete solar and galactic radiation is reaching the surface. Also the composition of the atmosphere is much different to Earth. The atmosphere is mainly consisting of CO_2, some N_2, Argon, and trace gases including some oxygen and water. The surface shows weak amounts of organics and is dominated by volcanic and/or phyllosilicate rich regolith. The water is mainly available in the form of ice. But in some areas, the activity of liquid water was supposed to be observed on the Martian surface (McEwen et al. 2014; Ojha et al. 2015; Dundas et al. 2017) what is explained as to be realized by a potential briny consistence. Even in some deep valleys or craters, the conditions are reaching a level above the triple point of about 6 mbar, where water could also be as well solid, liquid, and gaseous (Haberle et al. 2001).

Exactly these findings were used as fundament for all laboratory simulations including the simulations performed as pre-flight experiments of the space exposure experiments. But it should be clarified that it is always an approach to the real planetary conditions, because the lower gravity and the complete radiation spectra

Table 2.1 BIOMEX selected samples for spaceflight

Archaea	Methanogens	*Methanosarcina soligelidi* strain SMA-21 (terrestrial permafrost) (GFZ/AWI Potsdam, Germany)
Bacteria	Deinococcus	*Deinococcus radiodurans* wild type and crtI or crtB (nonpigmented) (DLR Cologne, Germany)
	Biofilm	Biofilm containing *Leptothrix, Pedomicrobium, Pseudomonas, Hyphomonas, Tetrasphaera* (TU Berlin, Germany)
	Cyanobacteria	Cyanobacterium *Nostoc* sp. strain CCCryo 231-06 (Fraunhofer IZI-BB Potsdam, Germany)
		Cyanobacterium *Gloeocapsa* OU-20 (Astrobiology Center Edinburgh, UK)
		Cyanobacterium *Chroococcidiopsis* sp. CCMEE 029 (Uni Roma, Italy)
Bacteria / Eukaryotes	Biofilm	Kombucha biofilm containing: Yeasts: *Saccharomyces ludwigii, Schizosaccharomyces pombe, Zygosaccharomyces rouxii, Zygosaccharomyces bailii, Brettanomyces bruxellensis;* Bacteria: *Paenibacillus* sp. IMBG221, *Acetobacter nitrogenifigens, Gluconacetobacter komhuchae* sp. nov., *Gluconacetobacter xylinum* (NAS Kiev, Ukraine)
Eukaryotes	Alga	Green alga *Sphaerocystis* sp. CCCryo 101-99 (Fraunhofer IZI-BB Potsdam, Germany)
	Lichens	*Circinaria gyrosa* (INTA Madrid, Spain)
		Buellia frigida (Antarctic lichen) (H-H-Uni Düsseldorf, Germany)
	Fungi	Cryptoendolithic Antarctic black fungus *Cryomyces antarcticus* CCFEE 515 (Uni Viterbo, Italy)
	Bryophytes	*Grimmia sessitana* (alpine samples) (Uni Potsdam, Germany)
		Marchantia polymorpha L. (Uni Potsdam, Germany)
Biomolecules	Pigments	Pigment Chlorophyll (H-H-Uni Düsseldorf, Germany)
		Pigment beta-Carotene (H-H-Uni Düsseldorf, Germany)
		Pigment Naringenin (H-H-Uni Düsseldorf, Germany)
		Pigment Quercitin (H-H-Uni Düsseldorf, Germany)
		Pigment Parietin (H-H-Uni Düsseldorf, Germany)
		Pigment Melanin (H-H-Uni Düsseldorf, Germany)
	Cell wall / Membrane components	Cellulose (H-H-Uni Düsseldorf, Germany)
		Chitin (H-H-Uni Düsseldorf, Germany)
Substrates/Minerals	substrate	Agar (as a substitute for Murein) (H-H-Uni Düsseldorf, Germ.)
	Lunar analog	Minerals lunar analog mixture (MfN Berlin, Germany)
	Mars analog	Minerals P-MRS: Early acidic Mars analog (Mixture of Fe_2O_3, montmorillonite, chamosite, kaolinite, siderite, hydromagnesite, quartz, gabbro, and dunite) (MfN Berlin, Germany)
		Minerals S-MRS: Late basic Mars analog (Mixture of hematite, goethite, gypsum, quartz, gabbro, dunite) (MfN Berlin, Germ.)
	Glass	Silica discs (glass) (Astrobiology Center Edinburgh, UK)

Green colored lines: photosynthetic active organisms

ranges as they are relevant on the surface of Mars are not to be realized simultaneously in laboratories on ground. A step further is to do those planetary simulation experiments directly in space, where extensive and costly simulation instruments are not needed to be applied to finally reach the real space and Mars-like environmental conditions. These needed parameters are all freely available in Low Earth Orbit and, for this reason, it was decided in the frame of the BIOMEX (*BIO*logy and *M*ars *EX*periment) project to investigate a significant amount of organisms of all three domains of the tree of life (see Table 2.1) directly on the EXOPOSE-R2 platform of the ISS.

It has to be clear that in LEO we are also just approaching the environmental conditions on the surface and the upper subsurface of Mars, because the magnetic field of Earth is still shielding against corpuscular radiation or in other orbit tracks additional proton and electron radiation of the radiation belt in LEO should be added. But calculations have shown that still in case of the solar irradiation, the sun-exposed samples are receiving in LEO much more Solar Constant hours compared to the surface of Mars (Earth: up to 1879 eSCh (estimated Solar Constant hours versus 1572 eSCh on Mars; data received from the LIFE (Lichen und Fungi Experiment) on EXPOSE-E of the COLUMBUS module in 2007–2008, Onofri et al. 2012; Rabbow et al. 2012)).

To get a feeling about how close the simulated conditions in LEO were approaching Mars-like conditions, the following list of the EXPOSE-E mission in relation to the LIFE experiment gives an overview (see Onofri et al. 2015):

1. Atmospheric composition and pressure reached in the compartments of the Mars simulation trays: 1.6% argon, 0.15% oxygen, 2.7% nitrogen, 370 ppm H_2O, 95 % CO_2 at a pressure of 10^3 Pa.
2. Cosmic radiation (maximum dose rate at the sample site: 368–27 µGy/day according to Berger et al. 2012). This dose rate was slightly higher than the 210–40 µGy/day measured by Hassler et al. (2014) in Gale Crater on Mars during the Mars Science Laboratory mission. In addition, the spectra are different; whereas on Mars galactic cosmic rays prevail, protons and electrons of the radiation belts must be added to the galactic radiation source in the orbit of the ISS (Dachev et al. 2012).
3. UV radiation (maximum 1572 solar constant hours of solar electromagnetic radiation at $\lambda > 200$ nm, resulting in a fluence at the sample site of 475 MJ/m^2 for 200 nm $< \lambda < 400$ nm UV, or 630 kJ/m^2 beneath a neutral density filter of 0.1% transmission). Because the Martian solar constant amounts to 45% of Earth's solar constant, the applied radiation would be equal to 3493 Mars solar constant hours.
4. Long-term exposure (559 days in operation, total time in space 583 days): this period corresponds to an exposure of nearly 1 Martian year (687 days).
5. Temperature was not actively controlled and oscillated between −21.7 °C and +42.9 °C with a one-time peak at 62 °C for a few hours (Rabbow et al. 2012). This range differed substantially from the temperatures on the surface of Mars, which can reach about 20 °C as a maximum at noon at the equator and −153 °C as a minimum at the poles.

Data from the EXPOSE-R2 mission with relevance to BIOMEX are showing the following Mars-like approach:

1. Atmospheric composition and pressure reached in the compartments of the Mars simulation trays: 95.55% CO_2, 2.70% N_2, 1.60% Ar, 0.15% O_2, ~370 ppm H_2O (Praxair Deutschland GmbH) at a pressure of 980 Pa.
2. Cosmic radiation (maximum dose rate at the sample site within 444 days: between 59 µGy/day and 77 µGy/day according to Dachev et al. 2017). This dose rate was slightly within the dose rate of 210–40 µGy/day measured by Hassler et al. (2014) in Gale Crater on Mars during the Mars Science Laboratory mission. In addition, the spectra are different; whereas on Mars galactic cosmic rays prevail, protons and electrons of the radiation belts must be added to the galactic radiation source in the orbit of the ISS (Dachev et al. 2012, 2017).

3. The calculated mean UV radiation fluence (means of the corresponding individually determined fields of view of the samples per compartment) of the biologically active wavelength range of 200–400 nm was 458 ± 32 MJ/m^2 for tray 1 and 492 ± 66 MJ/m^2 for tray 2 (Rabbow et al. 2012; Wadsworth et al. 2019). This means the radiation is approximately comparable to what was measured during the EXPOSE-E mission (see description in (3) above), what corresponds in the case of EXPOSE-R2 to about in minimum 3368 and in maximum about 3618 Mars solar constant hours.
4. Long-term exposure 509 days in operation: 62 days of which they were in the dark for outgassing, which was followed by 469 days of exposure to UV radiation (total time in space 531 days) This is slightly less than a Martian year (687 days).
5. Temperatures experienced on the ISS ranged from -20.9 to 58.0 °C without any individual temperature peak as during the EXPOSE-E mission (Rabbow et al. 2012). This range differed substantially from the temperatures on the surface of Mars, which can reach about 20 °C as a maximum at noon at the equator and -153 °C as a minimum at the poles. Just specific latitudes near to the Martian equator are approached.

It has to be mentioned that the slight differences between EXPOSE-E and EXPOSE-R2 measurements and calculations can be explained by different solar activity cycle what is also influencing the income of cosmic galactic rays.

2.2 Selection of Microorganisms for Habitability Studies

As described in Sect. 1.2, the different organisms (see Table 2.1) were selected due to their original Mars-analog habitats in selected Mars-analog field sites of the Arctic, Antarctica, and Alpine high altitudes (see Sect. 1.2.1). Species from all three domains of the tree of life as there are archaea, bacteria, and eukaryotes were chosen for further experiments on ground and in space. This chapter will just select some of the listed and investigated BIOMEX-relevant organisms mentioned in Table 2.1 as examples and are relating them to other chosen samples of previously performed space experiments. In some cases, samples were chosen as replicates to previous experiments performed on the ISS or on satellites. This was the case for the Antarctic fungus *Cryomyces antarcticus* (Onofri et al. 2008, 2012, 2015) and the Antarctic lichen *Circinaria gyrosa* (formerly classified as *Aspicilia fruticulosa* and afterwards reclassified according to Sohrabi et al. 2013, see also Sánchez et al. 2010; de la Torre et al. 2010). Both of these species had very limited space on the previous facilities and needed further replicates to get a better view and evaluation on their resistance to space and Mars-like conditions. *Cryomyces antarcticus* was also interesting to be further investigated because it was collected in the Mars-analog field sites of the Dry Valleys and pre-selection experiments have shown together with phylogenetically close relatives a certain degree of physiological activity during simulated Mars-like conditions in the lab what was even leading to protein

• Lichens survive and are photosynthetically active in Mars-like environments (de Vera et al. 2010, 2012, 2014, de la Torre et al. 2018)

• Methanogens survive and are active in Mars-like environments (Schirmack et al. 2014, Serrano et al. 2019)

• Fungi survive and produce proteins in Mars-like environments (Zakharova et al. 2014)

• Cyanobacteria survive Mars-like environments (de Vera et al. 2014)

Fig. 2.1 Pre-selection experiments at the Mars Simulation facility (MSF) at the Institute of Planetary Research of the German Aerospace Center (DLR) in Berlin: the results indicated that the chosen organisms were able to be active during the Mars simulation experiments—a good starting position to be further investigated in space

production (Zakharova et al. 2014; Fig. 2.1). Also both of those lichen and fungus species have interesting biominerals and biomolecules (see Chap. 1, Böttger et al. 2014; Pacelli et al. 2019, 2020).

Other species from the domain of archaea and bacteria were selected because they have shown also physiological activities during Mars simulation experiments in the Pre-selection phase. Besides the often deeply investigated radiation resistant and therefore Mars-relevant bacterium *Deinococcus radiodurance* (Leuko et al. 2017), the Siberian polar methanogen archaeon *Methanosarcina soligelidi* SMA-21 was chosen to be used for the space experiments. This archaeon was able to produce increasingly methane during Mars simulation, also a clear indication of its capability to grow under the investigated conditions (Schirmack et al. 2014; Serrano et al. 2019).

An interesting group of organisms are also photosynthesizing microorganisms, which could use the atmospheric Martian gas CO_2 and water to produce O_2 what should also have implications for future life supporting systems or terra forming issues on the red planet (Verseux et al. 2016). Therefore, a set of desert and Antarctic cyanobacteria and alga were chosen such as cryptoendolithic Chroococcidiopsis sp. CCMEE 029 of the Negev Desert in Israel and the Antarctic cyanobacterium *Nostoc* sp. strain CCCryo 231-06 as well as the Arctic alga *Sphaerocystis* sp. CCCryo 101-99.

Besides of the investigated cyanobacteria some of the Arctic and Antarctic lichens were also able to photosynthesize under Mars-like conditions (de la Torre Noetzel et al. 2018; de Vera et al. 2010, 2014a, b). The lichen specie *Xanthoria elegans* was an excellent example which has also been investigated during the

EXPOSE-E mission within the exposure experiment LIFE. It has shown photosynthetic activity during the performed Mars simulation experiments in the laboratory (de Vera et al. 2010) and survived also to about 80–90% the space exposure experiment (Brandt et al. 2014; de Vera 2012; Onofri et al. 2012). Because *X. elegans* is a lichen known from Alpine and Polar Regions, the follow-on selection for the BIOMEX project concentrated on cyanobacteria and lichens collected in such corresponding Mars-analog field sites. The question was to check, if such kind of significant resistance and adaptation potential is also valid for other species found in the same planetary analog field sites. Our choice as shown in Table 2.1 was to investigate *Buellia frigida* (Antarctic lichen). *B. frigida* is an endemic, crustose lichen that colonizes habitats of the maritime and continental Antarctica (Øvstedal and Lewis Smith 2001). The lichen grows on exposed rock surfaces and is well adapted to high fluxes of PAR, desiccation, and cold temperatures (Sadowsky and Ott 2012; Backhaus et al. 2014). The samples were collected in Antarctica, specifically in the North Victoria Land (74°38′S 164°13′E) during the GANOVEX X expedition in the Antarctic summer season 2009/2010 (Backhaus et al. 2019a, b). Further samples were in the form of biofilms. Whereas the Technische Universität Berlin has provided an Iron-bacteria dominated biofilm (*Leptothrix*, *Pedomicrobium*, *Pseudomonas*, *Hyphomonas*, *Tetrasphaera*) which could have some relevance for potential life on Mars using iron-rich resources on the surface of this planet, another biofilm was provided by the National Academy of Science (NAS) Kiev, which is well-known as probiotic culture named KOMBUCHA in drinks and yoghurts. The latter was originally classified as less relevant for Mars habitability studies. The focus of the KOMBUCHA studies was much more related to study the stability of this consortium of microorganisms to space and Mars-like conditions in reference to the use of such kind of probiotic nutrient as trigger and maintenance of the Astronaut's immune system in future long-term space missions (Podolich et al. 2017, 2019). Finally, also two species of Bryophytes (*Grimmia sessitana, Marchantia polymorpha* L.) of Alpine regions were tested to get a feeling, where the limits of survival on Mars should be reached (Huwe et al. 2019).

2.3 Sample Preparation

Except of the lichens and mosses, all samples were grown on glass, agar-, Lunar- or Mars-analog regolith simulant pellets before launch into space and final space exposure.

In this paragraph, a closer look is necessary on the preparation of the lichens and mosses, because the preparation was a bit different; this means, it was not possible to grow these organisms on the Mars-analog substrates in time (e.g. lichen: about 1 mm a year in max.).

With respect to the lichens, three types of samples were prepared for the experiments to be performed on the exposure platform EXPOSE-R2 at the ISS at the beginning of the BIOMEX project. Disks (12 mm diameter) of rock sample surfaces colonized in nature by the lichen *B. frigida* were carefully drilled from field collected colonized rock samples so that the underlying rock remained attached to the lichen. The base of the two other types of samples consisted of pellets of the two different Mars regolith simulants (as described in Böttger et al. 2014 and see Sect. 1.3). The Mars regolith powders were pressed to 12-mm-diameter disks. The P-MRS sample pellets represent early acidic Mars regolith and consisted of montmorillonite, chamosite, quartz, iron(III)-oxide, kaolinite, siderite, hydromagnesite, gabbro, and dunite (see de Vera et al. 2019), whereas the S-MRS sample pellets are similar in composition to late basic Mars regolith and consisted of gabbro, gypsum, dunite, hematite, goethite, and quartz (ibid). Four to five spots were selected on the top of each pressed pellet to serve as attachment points to which four to seven thallus fragments with sizes of about 2–3 mm^2/fragment were attached and except of the glue connecting point were also touching directly the Mars regolith simulants (Fig. 2.2a). The thallus fragments covered the artificial Mars regolith and were glued on one end in direct contact to each spot using RTV-S 691 silicone rubber (Wacker) glue (Fig. 2.2a), approved for aerospace applications by the European Space Research and Technology Centre (ESTEC). During the entire experiment, all three types of lichen samples were in a dry and anhydrobiotic state. All three setups of the samples (hereafter referred to as rock, P-Mars, and S-Mars samples) were sent to the DLR in Cologne, and they were subsequently integrated into a BIOMEX assigned tray of the EXPOSE-R2 facility for later exposure outside of the ISS. After return from the ISS, all samples were stored at −80 °C until further investigation (Backhaus et al. 2019a, b).

Bryophytes were prepared slightly different (Fig. 2.2b): three individual cushions of *Grimmia* gametophytes were separately washed with ddH$_2$O to remove rock particles and dust. The gametophytes from each cushion were then singled to plantlets (unbranched stems with leaves). 30 plantlets to 50 plantlets per cushion were reassembled to small cushions, cut to the same length (5 mm) and air dried under a laminar flow hood for 24 h to ensure comparability. These artificially produced cushions were then fixed to Mars regolith simulants (pellets of 11 mm diameter and 1 mm height) with space proof, two-component glue that is commonly used to glue material under real space conditions (Wacker RTV-S 691 A/B; Wacker Chemie AG, Germany, see also Huwe et al. 2019).

2.4 Pre-flight Tests: Habitability Test Results Under Simulated Mars-Like Conditions

The pre-flight habitability tests are separated pre-selection tests performed partly in the Mars simulation facility at the German Aerospace Center (DLR) at the Institute of Planetary Research in Berlin and partly as EVTs and SVT in the Microgravity User Support Center (MUSC) at the DLR Colgone (detailed description on EVTs and SVTs as well as MSG tests, see Sect. 1.4).

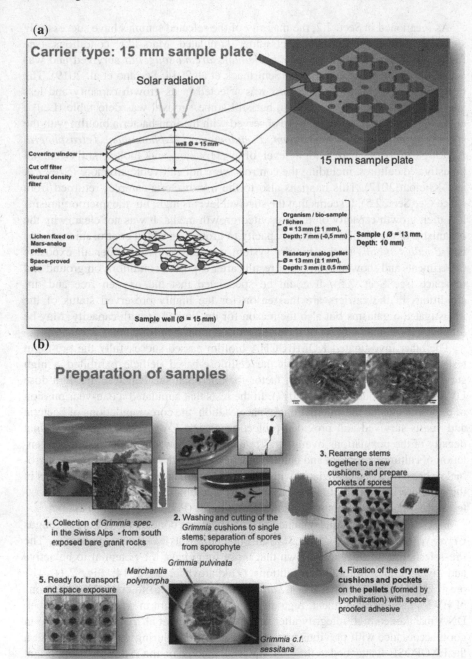

Fig. 2.2 (a) Sample preparation in the flight hardware (example with a lichen). (b) Preparation of Bryophytes on the Mars-analog pellet samples before launch into space

As mentioned in Sect. 2.2, the majority of the selected samples have successfully survived different pre-flight tests under simulated Mars-like conditions (see Fig. 2.1). The methanogen archaeon *Methanosarcina soligelidi* survived and was active under Mars-like conditions (Schirmack et al. 2014; Serrano et al. 2019). The bacterium *Deinococcus radiodurans* was affected in its growth capacity and less surviving the simulated conditions, but still some survival was detectable (Leuko et al. 2017). Intriguing results were observed with the iron-bacteria biofilm with the species *Leptothrix, Pedomicrobium, Pseudomonas, Hyphomonas, Tetrasphaera.* Although the cells had a high level of preservation and looked still vital, all investigated cultures, including the controls were not cultivable anymore (Szcwzyk and Kliefoth 2017). This happens also to the iron-bacteria samples returned from space (see Sect. 2.5). It seems that the survival level is high, but the microorganisms lost their growth capacity on the presented growth media. It was not clear, why the organisms of the biofilm lost their capacity to grow. The added control microorganism *Bacillus subtilis* to this biofilm system was still cultivable after all exposure experiments and showed even better results after exposure conditions on ground and in space (see Sect. 2.5). It could be speculated that the oxygen free and dry conditions in the carriers are the reason for the highly preserved status of the investigated organisms but also the reason for the limited growth capacity. May be that even the contamination of the samples might have an influence.

The other investigated KOMBUCHA biofilm passed successfully the pre-flight tests. KMC members that inhabit the cellulose-based pellicle exhibited a high survival rate. The critical limiting factor for microbial survival was the high-dose UV irradiation (Podolich et al. 2017). In the tests that simulated a one-year mission of exposure outside the International Space Station, the core populations of bacteria and yeasts survived and provided protection against UV; however, the microbial density of the populations overall was reduced, which was revealed by implementation of culture-dependent and culture-independent methods. A reduction of microbial richness was also associated with a lower accumulation of chemical elements in the cellulose-based pellicle film, produced by microbiota that survived in the post-test experiments, as compared to untreated cultures (Podolich et al. 2017).

In reference to the results obtained by investigations done on the Antarctic fungus *Cryomyces antarcticus* CCFEE 515, all Pre-Flight tests were very successful. The Pre-selection tests still have shown that those fungi have the potential to be active and living under Mars-like conditions (Zakharova et al. 2014; Fig. 2.1). The pre-flight tests have shown that *C. antarcticus* was able to tolerate the conditions of EVT and SVT experiment, regardless of the substratum in which it was grown. DNA maintained high integrity after treatments (Pacelli et al. 2019). This is also in good accordance with previous experiments performed during the LIFE project on the EXOPOSE-E attached to the COLUMBUS module on the ISS. This is important to emphasize, because the needed replicate pre-flight experiments for the ISS some years after the first ISS experiment supported previously obtained results (Onofri et al. 2012, 2015).

With respect to the lichens, it has to be emphasized that according to new replicate investigations within the BIOMEX project more detailed analysis were possible on the species *Circinaria gyrosa* (formerly named *Aspicilia fruticulosa*). It comes out that the lichen's photosynthetic activity was much more affected by radiation (48% activity) compared to samples in the dark areas only facing vacuum and Mars-like atmosphere (about 80% activity) (de la Torre Noetzel et al. 2020). These results are different to previously obtained pre-flight experiments with this lichen on ground particularly in reference to the space flight experiment BIOPAN 6 on the satellite FOTON M6 (de la Torre Noetzel et al. 2010). In the BIOPAN 6 experiment *C. gyrosa* maintained its photosynthetic activity nearly at 100%. It has to be emphasized that the satellite experiment was a short-term experiment in LEO (about 10 days) and, therefore, also the pre-flight experiments were done in very short time frames. But during the 1.5 years performed BIOMEX experiment other outcome of additionally performed investigations on this lichen species have also to be mentioned: polymerase chain reaction analyses confirmed that there was DNA damage in lichen exposed to harsh space and Mars-like environmental conditions, with ultraviolet radiation and also UV combined with simulated space vacuum what is causing the most damage (de la Torre Noetzel et al. 2020). This is in contrast to the results we got with another lichen species, *Xanthoria elegans*, in the long-term experiment LIFE (about 1.5 years in space), which was much more resistant to simulated Mars-like and space conditions (on ground and in space, see de Vera et al. 2010; Onofri et al. 2012). The other investigated lichen *Buelia frigida* showed promising results during the EVT and SVT phases. The results demonstrated that *B. frigida* is capable of surviving the conditions tested in EVT and SVT. The mycobiont showed lower average impairment of its viability than the photobiont (viability rates of >83% and >69%, respectively); and the lichen thallus suffered no significant damage in terms of thalline integrity and symbiotic contact (Meeßen et al. 2015).

The pre-flight test results on the Bryophyte *Grimmia sessitana* showed that severe UVR 200–400 nm irradiation, respectively, was the only stressor with a negative impact on the vitality (37% (terrestrial atmosphere) or 36% reduction (space- and Mars-like atmospheres)) what is based on results obtained through investigations on the photosynthetic activity. With every exposure to UV, the vitality of the bryophytes dropped by 6%. No effect was found, however, by any other stressor (Huwe et al. 2019).

The results on the listed Cyanobacterium *Chroococcidiopsis* have shown reduced growth capacity under simulated Mars and space like conditions, if embedded in the Mars regolith simulants. No growth was obvious to samples not embedded in the Mars regolith (Billi et al. 2019a, b).

Some short notice should be made that the investigated green alga *Sphaerocystis* sp. CCCryo 101-99 and the cyanobacterium *Nostoc* sp. strain CCCryo 231-06 provided by the Fraunhofer-IZI-BB Potsdam and listed in Table 2.1 were able to grow after the Pre-selection tests (further post-flight analysis are going on and some articles are currently in preparation).

2.5 Results After Sample Return and Post-Flight Investigations

Whereas the post-flight results on the methanogen archaeon *Methanosarcina soligelidi*, the cyanobacterium *Nostoc* sp. strain CCCryo 231-06, the alga *Sphaerocystis* sp. CCCryo 101-99 and the Bryophytes (*Grimmia sessitana, Marchantia polymorpha* L.) are pending the following post-flight results could be listed here in reference to the investigated organisms:

- *Deinococcus radiodurans*: was affected in its growth capacity and less surviving the space and Mars-like conditions on the ISS, but still some survival was detectable (Leuko et al. 2017).
- Biofilm of iron bacteria (*Leptothrix, Pedomicrobium, Pseudomonas, Hyphomonas, Tetrasphaera*): although the cells had a high level of preservation and looked still vital, all investigated cultures, including the controls were not cultivable anymore (Szewzyk and Kliefoth 2017).
- *Cyanobacterium Chroococcidiopsis* sp. CCMEE 029: Survival occurred only for dried cells (4–5 cell layers thick) mixed with the martian regolith simulants P-MRS and S-MRS, and viability was only maintained for a few hours in references to samples exposed to space with a total UV (wavelength from 200 to 400 nm) radiation dose of 492 MJ/m^2 (attenuated by 0.1% neutral density filters) and 0.5 Gy of ionizing radiation.
- Kombucha Biofilm: post-exposure analysis demonstrated that growth was observed of both the bacterial and yeast members of the KMC community after 60 days of incubation, whereas growth was detected after 2 days in the initial KMC. The KMC that was exposed to extraterrestrial UV radiation showed degradation of DNA, alteration in the composition and structure of the cellular membranes, and an inhibition of cellulose synthesis. In the "space dark control" (exposed to LEO conditions without the UV radiation), the diversity of the microorganisms that survived in the biofilm was reduced compared with the ground-based controls. This was accompanied by structural dissimilarities in the extracellular membrane vesicles. After a series of sub-culturing, the revived communities restored partially their structure and associated activities (Podolich et al. 2019).
- *Circinaria gyrosa*: samples exposed to UV radiation >110 nm and to space vacuum exhibited the lowest photosynthetic activity (11.9–30.2%), whereas those exposed to Mars-like UV radiation and to a Mars-like atmosphere composition and pressure showed a higher photosynthetic activity range (40–48.5%) approaching the results of the pre-flight experiments (see Sect. 2.4). The Polymerase chain reaction analyses confirmed that there was DNA damage in lichen exposed to harsh space and Mars-like environmental conditions, with ultraviolet radiation and also UV combined with simulated space vacuum what is causing the most damage (de la Torre Noetzel et al. 2020).

- *Buellia frigida*: The results on the Antarctic lichen were the most intriguing results. Although this lichen was collected in a Mars-analog field site in Antarctica and was classified through EVT and SVT tests as a promising candidate to resist Space and Mars-like conditions in orbit (Meeßen et al. 2015, Sect. 2.4), this lichen had the worse results in the post-flight experiments. The mortality rates were up to 100% for the algal symbiont and up to 97.8% for the fungal symbiont. In contrast, the lichen symbiont controls exhibited mortality rates of 10.3% up to 31.9% for the algal symbiont and 14.5% for the fungal symbiont (Backhaus et al. 2019a, b)
- *Cryomyces antarcticus*: The results demonstrate that *C. antarcticus* was able to tolerate the combined stress of different extraterrestrial substrates, space, and simulated Mars-like conditions in terms of survival, DNA, and ultrastructural stability (Onofri et al. 2019). One example: the survival, expressed as CFUs obtained culturing cells of C. antarcticus, once back from space was very high, ranging from 40% to 80% in fully irradiated samples, investigated in both space and Mars-like conditions. Survival of in-flight dark samples in vacuum and in Mars atmosphere was also elevated, but the latter was apparently more harmful (Onofri et al. 2019).

2.6 Conclusions on the Habitability of Mars and Outlook

According to the presented results obtained before and after space exposure on EXPOSE-R2/ISS during the BIOMEX experiment, it was clear that for many of the investigated organisms we could not exclude that they are able to survive or even to live under Mars-like conditions. Mars seems to be habitable for some species of cyanobacteria and fungi (see Table 2.2) at least in some niches close to the Martian surface or in the subsurface. It may also be habitable for some species of archaea, single bacteria, bacteria, and eukaryotes containing biofilms, alga, and lichens. But we have to be careful to generalize. It seems that a few species of the domains of the tree of life are able to resist or to be active under the tested conditions of pre- and post-flight conditions. If we consider the big group of lichens, it was obvious that some lichens of polar and alpine habitats were able to survive and be active under Mars-like condition, such as it was observed by analysis on the highly resistant lichen *X. elegans* in the lab (de Vera et al. 2010) and in space during the EXOPSE-E mission (Onofri et al. 2012). Also it is not mandatory that organisms collected in Mars-analog field sites are the prime objects to be studied further under simulated planetary conditions, as was obvious by the results gained with the non-resistant Antarctic lichen *B. frigida* and the partly resistant Kombucha biofilm, which was not selected in a Mars-like environment. Therefore, no rule could be formulated that makes it prerequisite to collect mainly samples in Mars-analog field sites. There is just a tendency to find more adapted species in those planetary analog field sites compared to other environments, what was visible by the results obtained with the Antarctic fungus *C. antarcticus* and the Antarctic cyanobacterium *Nostoc* sp. strain

Table 2.2 Habitability of Mars to the investigated organisms in reference to the pre- and post-flight results

Domain	Specie group	Investigated species	Is Mars habitable? Yes	May be	No
Archaea	Isolate	*Methanosarcina soligelidi*		x[a]	
Bacteria	Isolate	*Deinococcus radiodurans*		x	
	Biofilm of iron bacteria	Biofilm containing *Leptothrix*, *Pedomicrobium*, *Pseudomonas*, *Hyphomonas*, *Tetrasphaera*			x
	Cyanobacteria	*Nostoc* sp. strain CCCryo 231-06	x[a]		
		Chroococcidiopsis sp. CCMEE 029	x		
Bacteria/ eukaryotes	Biofilm	Kombucha biofilm		x	
Eukaryotes	Alga	Green alga *Sphaerocystis* sp. CCCryo 101-99 (Fraunhofer IZI-BB Potsdam, Germany)		x[a]	
	Lichens	*Circinaria gyrosa* (INTA Madrid, Spain)		x	
		Buellia frigida (Antarctic lichen) (H-H-Uni Düsseldorf, Germany)			x
	Fungi	Cryptoendolithic Antarctic black fungus *Cryomyces antarcticus* CCFEE 515 (Uni Viterbo, Italy)	x		
	Bryophytes	*Grimmia sessitana* (alpine samples) (Uni Potsdam, Germany)			x[a]

[a]Personal communications by the Co-investigators of BIOMEX

CCCryo 231-06 as well as the cyanobacterium *Chroococcidiopsis* sp. CCMEE 029. Because further investigations of the BIOMEX project are ongoing, we should wait for definitive final conclusions on the question, how habitable Mars could be for terrestrial organisms. These results have also implications on Planetary Protection. If numerous terrestrial organisms are able to survive or even to be active on the planet Mars, it is necessary to sterilize the probes and their instruments before landing on this planet.

The value of replicates should also be emphasized for future space experiments. The amount of sample holders in the available facilities is limited. Therefore, the necessity to do replicates in current or future space exposure experiments for comparison to the previously performed space exposure experiments is always a scientific must.

Samples which have been analyzed by one method and was not exhausted during these investigations should be stored for further analysis. This is also the reason, why investigations after space exposure are ongoing even after months and years after the sample return.

As it is valid for all experiments, always an improvement of sample preparation, methods and operations is needed and ongoing. By the still performed space experiments we have to learn, how to avoid contamination as it occurred for the experiments on the iron-bacteria biofilm. We have to adapt investigations to newly developed methods and should benefit of the modern investigation procedures in the next planned space experiments in LEO or even beyond without forgetting to apply the still well established and standardized analysis methods for comparability and reproducibility.

2.6.1   Limits of Life Above Earth's Atmosphere?

The experiments performed on the ISS were not just providing results to answer the questions on the habitability of another planet than Earth or on the detectability of life in extraterrestrial environments. By these experiments, we are able to get also knowledge on the general limits of life. It seems that according to the biological experiments performed in the last decades on the ISS and on some satellites we could carefully conclude that the open space above our Earth's atmosphere is not a real limit for life. Organisms in the dry and dormant state and if protected by covering natural or artificial material should survive the conditions and is even able to be reactivated after years may be after even millions of years. From a biological point of view, the origin and further evolution of life in space or on other planets and moons in the solar system and beyond is no longer questionable. The proof is also obvious: humankind made its step into space together with its microbiomes and is still living out of the Earth on the ISS (Fig. 2.3).

Fig. 2.3 EXPOSE-R2 filled with samples installed on the Russian Zvezda module on the ISS by astronauts (picture provided by ESA/Roscosmos)

2.6.2   The Likelihood of Lithopanspermia

It should be mentioned that since many years of space research on the ISS the obtained results on the exposure platforms EXPOSE-E, EXPOSE-R, and EXPOSE-R2 have also provided a lot of information on the capacity of terrestrial organisms to survive a space travel. Although this space travel is limited to about a few months to a few years in LEO, still knowledge was gained on the likelihood of a successful interplanetary transfer of life by these artificially caused space travels. By nature such kind of space travel of life could be initiated through impact events caused by asteroids or comets on a planet which is colonized by life and where colonized ejecta in the spallation zone could reach by its velocity the open space and an orbit which allows a travel to another planet (Arrhenius 1903, 1908; Mastrapa et al. 2001; Stöffler et al. 2007). Often such an impact event in the Earth-Mars-System served as a potential reference for a successful lithopanspermia between both planets (Horneck et al. 2008; Stöffler et al. 2007), because we still have evidence on such really occurring material transfer events through the discoveries made by more than 40 Martian meteorites discovered on Earth (Melosh 1984; Vickery and Melosh 1987; Weiss et al. 2000; Nyquist et al. 2001; Head et al. 2002; Artemieva and Ivanov 2004; Fritz 2005). It is clear that the time of space exposure in LEO is limited and deep space conditions might look much different to the LEO environment. But with space exposure experiments we are approaching those conditions and could get a feeling, how the investigated organisms do behave during the limited so-called long-term space exposure. Because a significant number of different species are able to survive these conditions and also the samples tested in the present BIOMEX project shows repetitively successful survival even with different samples compared to previously performed space exposure experiments, we could conclude that the likelihood of lithopanspermia in the solar system and specifically in the Earth-Mars trajectory is not zero (Horneck et al. 2008; Onofri et al. 2019; Onofri et al. 2012; Stöffler et al. 2007 and others).

2.6.3   Potential Use of Terrestrial Life on Mars

Based on the obtained results through the research done with life in space, also interesting application opportunities of some resistant organisms or even biofilms like the investigated Kombucha biofilm could play a role in triggering the immune system and serving as food for astronauts during long manned space missions. Other organisms, which are able to photosynthesize such as the tested cyanobacteria or alga could even be used for life supporting systems, because they are able to produce oxygen or fix nitrogen what is absolutely needed for successful space agriculture and food production with plants (Verseux et al. 2016). Another open question should also be further investigated: could photosynthesizing organisms be applied for terraforming e.g. Mars? This means that Astrobiology operations on the ISS are therefore directly also influencing the disciplines of terrestrial and potentially extra-terrestrial agricultures, life sciences, and space medicine. An interaction through all space-related biology fields is highly appreciated for the gain of maximum benefits to our planet Earth, to space, and to humankind.

2.6.4 The Role of the ISS to Find Life on Icy Moons

Future space experiments should not forget the use of further space exposure platforms on the ISS. Because the space missions in the last decades have also revolutionized our understanding of the Jovian and Saturnian satellites, a new astrobiological field of interest just started a few years ago: it is the research of icy ocean worlds. But what does this field of interest had to do with the exposure experiments on the ISS? The answer is clear: since we know that the moon of Saturn named Enceladus is constantly ejecting organic rich water, gas, silicate rich dust and even complex organics through plumes coming from ocean through the open cracks in the icy crust (Waite Jr et al. 2009), potentially also microorganisms or the building blocks of life could be discovered on this icy world. The same is valid for the Jovian moon Europa where sporadically such ejecta events are happening (Roth et al. 2014). This means that in future, we should do exposure investigations with organisms which are populating our oceans. Such exposure experiments on the ISS or even in orbit around the Moon or directly on the surface of the Moon performed with those organisms would mimic the plume events, where potential life could come directly in contact with the space environment. If we study our terrestrial ocean-related organisms in the space environment, we would be able to know, if those organisms still remain intact and could be detected on the cell level. We should also be able to give answers on the habitability status of these icy moons as well as on the likelihood of a lithopanspermia scenario between the outer and the inner solar system and in particular between those icy worlds and our home planet Earth. One of the proposed space exposure experiments was successfully selected by ESA to be realized in 2024: it is the mission BioSigN (see Sect. 1.9.2). I am looking forward to this new promising space experiment with certainly fascinating new findings.

References

Arrhenius S (1903) Die Verbreitung des Lebens im Weltraum. Umschau 7:481–485
Arrhenius S (1908) Worlds in the making: the evolution of the universe. Harper & Row, New York
Artemieva NA, Ivanov BA (2004) Launch of Martian meteorites in oblique impacts. Icarus 171:183–196
Backhaus T, de la Torre R, Lyhme K, de Vera J-P, Meeßen J (2014) Desiccation and low temperature attenuate the effect of UVC254nm in the photobiont of the astrobiologically relevant lichens Circinaria gyrosa and Buellia frigida. Int J Astrobiol 14:479–488

Backhaus T, Meeßen J, Demets R, de Vera J-P, Ott S (2019a) Characterization of viability of the lichen Buellia frigida after 1.5 years in space on the International Space Station. Astrobiology 19 (2):233–241. https://doi.org/10.1089/ast.2018.1894

Backhaus T, Meeßen J, Demets R, de Vera J-PP, Ott S (2019b) DNA damage of the lichen Buellia frigida after 1.5 years in space using Randomly Amplified Polymorphic DNA (RAPD) technique. Planet Space Sci 177:104687. https://doi.org/10.1016/j.pss.2019.07.002

Berger T, Hajek M, Bilski P, Körner C, Vanhavere F, Reitz G (2012) Cosmic radiation exposure of biological test systems during the EXPOSE-E mission. Astrobiology 12:387–392

Billi D, Verseux C, Fagliarone C, Napoli A, Baqué M, de Vera J-P (2019a) A desert cyanobacterium under simulated mars-like conditions in low earth orbit: implications for the habitability of Mars. Astrobiology 19(2):158–169

Billi D, Mosca C, Fagliarone C, Napoli A, Verseux C, Baqué M, de Vera J-P (2019b) Exposure to low Earth orbit of an extreme-tolerant cyanobacterium as a contribution to lunar astrobiology activities. Int J Astrobiol 2019:1–8. https://doi.org/10.1017/S1473550419000168

Böttger U, de la Torre R, Frias J-M, Rull F, Meessen J, Sánchez Íñigo FJ, Hübers H-W, de Vera JP (2014) Raman spectroscopic analysis of the oxalate producing extremophile Circinaria Gyrosa. Int J Astrobiol 13(1):19–27

Brandt A, de Vera J-P, Onofri S, Ott S (2014) Viability of the lichen Xanthoria elegans and its symbionts after 18 months of space exposure and simulated Mars conditions on the ISS. Int J Astrobiol. https://doi.org/10.1017/S1473550414000214

Dachev T, Horneck G, Häder D-P, Schuster M, Richter P, Lebert M, Demets R (2012) Time profile of cosmic radiation exposure during the EXPOSE-E mission: the R3DE instrument. Astrobiology 12:403–411

Dachev TP, Bankov NG, Tomov BT, Matviichuk YN, Dimitrov PG, Häder D-P, Horneck G (2017) Overview of the ISS radiation environment observed during the ESA EXPOSE-R2 mission in 2014–2016. Space Weather 15:1475–1489. https://doi.org/10.1002/2016SW001580

de la Torre Noetzel R, Miller AZ, de la Rosa JM, Pacelli C, Onofri S, García Sancho L, Cubero B, Lorek A, Wolter D, de Vera JP (2018) Cellular responses of the lichen Circinaria gyrosa in Mars-like conditions. Front Microbiol 9:308. https://doi.org/10.3389/fmicb.2018.00308

de la Torre Noetzel R, Ortega Garcia MV, Zélia Miller A, Bassy O, Granja C, Cuberto B, Jordao L, Martinez Frías J, Rabbow E, Backhaus T, Ott S, Garcia Sancho L, de Vera JPP (2020) Lichen vitality after a space flight on board the EXPOSE-R2 facility outside the International Space Station: results of the biology and Mars experiment. Astrobiology 20(5):583–600

de la Torre R, Sancho LG, Horneck G, Ascaso C, de los Ríos A, Olsson-Francis K, Cockell CS, Rettberg P, Berger T, de Vera JPP, Ott S, Frías JM, Wierzchos J, Reina M, Pintado A, Demets R (2010) Survival of lichens and bacteria exposed to outer space conditions – results of the Lithopanspermia experiments. Icarus 208:735–748

de Vera J-P (2012) Lichens as survivors in space and on Mars. Fungal Ecol 5:472–479

de Vera JP, Möhlmann D, Butina F, Lorek A, Wernecke R, Ott S (2010) Survival potential and photosynthetic activity of lichens under Mars-like conditions: a laboratory study. Astrobiology 10(2):215–227

de Vera J-P, Schulze-Makuch D, Khan A, Lorek A, Koncz A, Möhlmann D, Spohn T (2014a) Adaptation of an Antarctic lichen to Martian niche conditions can occur within 34 days. Planet Space Sci 98:182–190

de Vera J-P, Dulai S, Kereszturi A, Koncz A, Lorek A, Möhlmann D, Marschall M, Pocs T (2014b) Results on the survival of cryptobiotic cyanobacteria samples after exposure to Mars-like environmental conditions. Int J Astrobiol 13(1):35–44

de Vera et al (2019) Limits of life and the habitability of Mars: the ESA space experiment BIOMEX on the ISS. Astrobiology 19(2):145–157

Dundas CM, McEwen AS, Chojnacki M, Milazzo MP, Byrne S, McElwaine JN, Ursoet A (2017) Granular flows at recurring slope lineae on Mars indicate a limited role for liquid water. Nat Geosci 10:903–907. https://doi.org/10.1038/s41561-017-0012-5

Fritz J (2005) Aufbruch vom Mars: Petrographie und Stoßwellenmetamorphose von Marsmeteoriten. Mathematisch-Naturwissenschaftliche Fakultät I. Humboldt-Universität zu Berlin, Berlin, p 138

Haberle RM, McKay CP, Schaeffer J, Cabrol NA, Grin EA, Zent AP, Quinn R (2001) On the possibility of liquid water on present-day Mars. J Geophys Res 106(10):23317–23326

Hassler DM, Zeitlin C, Wimmer-Schweingruber RF, Ehresmann B, Rafkin S, Eigenbrode JL, Brinza DE, Weigle G, Böttcher S, Böhm E, Burmeister S, Guo J, Köhler J, Martin C, Reitz G, Cucinotta FA, Kim MH, Grinspoon D, Bullock MA, Posner A, Gómez-Elvira J, Vasavada A, Grotzinger JP, MSL Science Team (2014) Mars' surface radiation environment measured with the Mars Science Laboratory's Curiosity rover. Science 343:1244797. https://doi.org/10.1126/science.1244797

Head JN, Melosh HJ, Ivanov BA (2002) Martian meteorite launch: high-speed ejecta from small craters. Science 298:1752–1756

Horneck G, Stöffler D, Ott S, Hornemann U, Cockell CS, Möller R, Meyer C, de Vera JP, Fritz J, Schade S, Artemieva N (2008) Microbial rock inhabitants survive hypervelocity impacts on mars like host planets: first phase of lithopanspermia experimentally tested. Astrobiology 8 (1):17–44

Huwe B, Fiedler A, Moritz S, Rabbow E, de Vera JP, Joshi J (2019) Mosses in low earth orbit: implications for the limits of life and the habitability of Mars. Astrobiology 19(2):221–232

Jensen LL, Merrison J, Hansen AA, Mikkelsen KA, Kristoffersen T, Nørnberg P, Lomstein BA, Finster K (2008) A facility for long-term mars simulation experiments: the mars environmental simulation chamber (MESCH). Astrobiology 8(3):537–548. https://doi.org/10.1089/ast.2006.0092

Leuko S, Bohmeier M, Hanke F, Böttger U, Rabbow E, Parpart A, Rettberg P, de Vera J-PP (2017) On the stability of deinoxanthin exposed to mars conditions during a long-term space mission and implications for biomarker detection on other planets. Front Microbiol 8:1680. https://doi.org/10.3389/fmicb.2017.01680

Mastrapa RME, Glanzberg H, Head JN, Melosh HJ, Nicholson WL (2001) Survival of bacteria exposed to extreme acceleration: implications for panspermia. Earth Planet Sci Lett 189:1–8

McEwen AS, Dundas CM, Mattson SS, Toigo AD, Ojha L, Wray JJ, Chojnacki M, Byrne S, Murchie SL, Thomas N (2014) Recurring slope lineae in equatorial regions of Mars. Nat Geosci 7:53–58

Meeßen J, Wuthenow P, Schille P, Rabbow R, de Vera JPP, Ott S (2015) Resistance of the lichen Buellia frigida to simulated space conditions during the preflight tests for BIOMEX – viability assay and morphological stability. Astrobiology 15(8):601–615

Melosh HJ (1984) Impact ejection, spallation, and the origin of meteorites. Icarus 59:234–260

Nyquist LE, Bogard DD, Shih C-Y, Greshake A, Stöffler D, Eugster O (2001) Ages and geological histories of Martian meteorites. Space Sci Rev 96:105–164

Ojha L, Wilhelm MB, Murchie SL, McEwen AS, Wray JJ, Hanley J, Massé M, Chojnacki M (2015) Spectral evidence for hydrated salts in recurring slope lineae on Mars. Nat Geosci 8:829–832

Onofri S, Barreca D, Agnoletti A, Rabbow E, Horneck G, de Vera JPP, Selbmann L, Zucconi L, Hatton J (2008) Resistance of Antarctic black fungi and cryptoendolithic communities to simulated space and Mars conditions. Stud Mycol 61(1):99–109

Onofri S, de la Torre R, de Vera JP, Ott S, Zucconi L, Selbmann L, Scalzi G, Venkateswaran K, Rabbow E, Horneck G (2012) Survival of rock-colonizing organisms after 1.5 years in outer space. Astrobiology 12(5):508–516

Onofri S, de Vera J-P, Zucconi L, Selbmann L, Scalzi G, Venkateswaran KJ, Rabbow E, de la Torre R, Horneck G (2015) Survival of Antarctic cryptoendolithic fungi in simulated Martian conditions on board the international space station. Astrobiology 15(12):1052–1105

Onofri S, Selbmann L, Pacelli C, Zucconi L, Rabbow E, de Vera J-P (2019) Survival, DNA, and ultrastructural integrity of a cryptoendolithic Antarctic fungus in mars and lunar rock analogues exposed outside the international space station. Astrobiology 19(2):170–182. https://doi.org/10.1089/ast.2017.1728

Øvstedal DO, Lewis Smith RI (2001) Lichens of Antarctica and South Georgia. A guide to their identification and ecology. Cambridge University Press, Cambridge, pp 361–363

Pacelli C, Selbmann L, Zucconi L, Coleine C, de Vera J-P, Rabbow E, Böttger U, Dadachova E, Onofri S (2019) Responses of the black fungus Cryomyces antarcticus to simulated mars and

space conditions on rock analogues. Astrobiology 19(1):209–220. https://doi.org/10.1089/ast.
2016.1631
Pacelli C, Cassaro A, Baqué M, Selbmann L, Zucconi L, Maturilli A, Botta L, Saladino R,
Böttger U, Demets R, Rabbow E, de Vera J-P, Onofri S (2020) Fungal biomarkers are detectable
in Martian rock-analogues after space exposure: implications for the search of life on Mars. Nat
Astron (under review)
Podolich O, Zaets I, Kukharenko O, Orlovska I, Reva O, Khirunenko L, Sosnin M, Haidak A,
Shpylova S, Rabbow E, Skoryk M, Kremenskoy M, Demets R, Kozyrovska N, de Vera J-P
(2017) Kombucha multimicrobial community under simulated spaceflight and Martian condi-
tions. Astrobiology 17:459–469
Podolich O, Kukharenko O, Haidak A, Zaets I, Zaika L, Storozhuk O, Palchikovska L, Orlovska I,
Reva O, Borisova T, Khirunenko L, Sosnin M, Rabbow E, Kravchenko V, Skoryk M,
Kremenskoy M, Demets R, Olsson-Francis K, Kozyrovska N, de Vera J-PP (2019)
Multimicrobial Kombucha culture tolerates mars-like conditions simulated on low-earth orbit.
Astrobiology 19(2):183–196
Rabbow E, Rettberg P, Barczyk S, Bohmeier M, Parpart A, Panitz C, Horneck G, von Heise-
Rotenburg R, Hoppenbrouwers T, Willnecker R et al (2012) EXPOSE-E: an ESA astrobiology
mission 1.5 years in space. Astrobiology 12:374–386
Robledo-Martinez A, Sobral H, Ruiz-Meza A (2012) Electrical discharges as a possible source of
methane on mars: lab simulation. Geophys Res Lett 39:L17202. https://doi.org/10.1029/
2012GL053255
Roth L, Saur J, Retherford KD, Strobel DF, Feldman PD, McGrath MA, Nimmo F (2014) Transient
water vapor at Europa's south pole. Science 343:171–174
Sadowsky A, Ott S (2012) Photosynthetic symbionts in Antarctic terrestrial ecosystems: the
physiological response of lichen photobionts to drought and cold. Symbiosis 58:81–90
Sánchez FJ, de la Torre R, Sancho L, Mateo-Martí E, Martínez-Frías J, Horneck G (2010) Aspicilia
fruticulosa: one of the most resistant organisms to outer space conditions and Mars simulated
environment. Orig Life Evol Biosph 40(6):546. https://doi.org/10.1007/s11084-010-9215-0
Schirmack J, Böhm M, Brauer C, Löhmannsröben H-G, de Vera J-P, Möhlmann D, Wagner D
(2014) Laser spectroscopic real time measurements of methanogenic activity under simulated
Martian subsurface analogue conditions. Planet Space Sci 98:198–204. https://doi.org/10.1016/
j.pss.2013.08.019
Serrano P, Alawi M, de Vera J-P, Wagner D (2019) Response of methanogenic archaea from
Siberian permafrost and non-permafrost environments to simulated Mars-like desiccation and
the presence of perchlorate. Astrobiology 19(2):197–208
Sohrabi M, Stenroos S, Myllys L, Søchting U, Ahti T, Hyvönen J (2013) Phylogeny and taxonomy
of the 'manna lichens'. Mycol Prog 12:231–269
Stöffler D, Horneck G, Ott S, Hornemann U, Cockell CS, Möller R, Meyer C, de Vera J-P, Fritz J,
Artemieva NA (2007) Experimental evidence for the impact ejection of viable microorganisms
from Mars-like planets. Icarus 186:585–588
Szewzyk U, Kliefoth M (2017) Untersuchungen zum Überleben und zur Aktivität von
Eisenbakterien unter Mars-ähnlichen Bedingungen auf der ISS und im Labor.
Abschlußbericht/Final Report and das DLR/BMWi Förderkennzeichen: 50WB1513, Institut
für Technischen Umweltschutz, Fachgebiet Umweltmikrobiologie, Technische Universität Ber-
lin, pp 1–45
Verseux C, Baqué M, Lehto K, de Vera J-PP, Rothschild LJ, Billi D (2016) Sustainable life support
on Mars – the potential roles of cyanobacteria. Int J Astrobiol 15(1):65–92
Vickery AM, Melosh HJ (1987) The large crater origin of SNC meteorites. Science 237:738–743
Wadsworth J, Rettberg P, Cockell CS (2019) Aggregated cell masses provide protection against
space extremes and a microhabitat for hitchhiking co-inhabitants. Astrobiology 19
(8):995–1007. https://doi.org/10.1089/ast.2018.1924
Waite JH Jr, Lewis WS, Magee BA, Lunine JI, McKinnon WB, Glein CR, Mousis O, Young DT,
Brockwell T, Westlake J, Nguyen MJ, Teolis BD, Niemann HB, McNutt Jr RL, Perry M, Ip WH

(2009) Liquid water on Enceladus from observations of ammonia and 40Ar in the plume. Nature 460:487–490

Weiss BP, Kirschvink JL, Baudenbacher FJ, Vali H, Peters NT, Macdonald FA, Wikswo JP (2000) A low temperature transfer of ALH84001 from Mars to Earth. Science 290:791–795

Zakharova K, Marzban G, de Vera J-P, Lorek A, Sterflinger K (2014) Protein patterns of black fungi under simulated Mars-like conditions. Sci Rep 4:5114. https://doi.org/10.1038/srep05114

Printed in the United States
By Bookmasters